LIFE THROUGH TIME

LIFE THROUGH TIME

HAROLD L. LEVIN

Washington University

Library of Congress Catalog Card Number: 74–17606

ISBN 0–697–05034–3

Second Printing, 1976

Printed in the United States of America

To my wife, Kay Helen,
who gave persistent encouragement,
I inscribe this book.

Contents

Preface

This book surveys the pageant of life as revealed by fossils over approximately three billion years. It is a book intended for those undergraduates who do not necessarily intend to specialize in a geologic science, but who desire, on aesthetic grounds, to obtain an understanding of man's biological background. The book is an outgrowth of a course taught at Washington University titled, "Life of the Geologic Past." Students who enroll in the class have a wide spectrum of interests, ranging from literature and philosophy to architecture and anthropology. Geology enthusiasts are also represented. All are there primarily because they recognize that knowledge of evolution and prehistoric life should become a part of their intellectual equipment. This kind of audience happens to find vertebrates more interesting than other organisms; therefore, animals with backbones receive emphasis in this book.

It is my hope that this small volume will foster other courses similar to "Life of the Geologic Past." Such offerings often bring students not particularly attuned to science into geology departments and provide them with the opportunity to examine the rewarding and fascinating world of fossils.

Life Through Time could not have been written without extensive recourse to the important contributions of those who have been in the forefront of paleontology for the past several decades. I am especially grateful for the instruction I have received from the volumes of E. H. Colbert and A. S. Romer. Although I prepared the majority of the illustrations, additional excellent renditions, recognizable by their distinctive styles, were provided by Miss Lynn Walters and Mrs. Evie Hill. Mrs. Margaret Bewig assisted me in the typing. Special thanks are due Mr. Raymond C. Deveaux for his editorial expertise and encouragement and to the students in "Life of the Geologic Past" who challenged me to write this book.

LIFE THROUGH TIME

1 Once Upon A Time

The search for a fossil may be considered at least as rational as the pursuit of a hare.
—William Smith (1769–1839)

INTRODUCTION

One of the unique characteristics of the primate known as man is his ceaseless questioning. Philosophers, scientists, and artists down through the centuries have asked, "Who are we, where did we come from, whither are we bound?" The ultimate answers to these questions still evade us, but partial answers are within our grasp. In this book, we shall consider the great chain of events which led from the enigmatic "half-life" of an ancient organic molecule, to the vast array of animals that people our planet. The basis for our study will not be found in the archives of libraries, but in the traces of once living things which seem to gaze outward at us from the rocks and ask our understanding.

TRACES OF ANCIENT LIFE

Little would be known about the history of life if it were not for those professional grave robbers known as paleontologists. It is they who disinter the bones. It is they who interpret the lively, petrified, intriguing traces of former life. Those traces are called *fossils*. Fossils are the natural hieroglyphics of the paleontologic record.

Frequently in the past, organisms were covered by the rain of sediment that falls continuously onto the seafloors or is blown about by the wind. On occasion, animals and plants were immersed in tar or quicksand, or even trapped in lava flows or a rain of volcanic ash. For preservation, a quick burial is important. In addition, if the animal happened to have a mineralized skeleton, all the better. After death, the flesh and other soft tissue is ordinarily consumed by bacteria and scavengers. The skeleton, be it of outside shell or inside bone, will be left to petrify in a gradually hardening matrix of sediment. The process is called *petrifaction,* and it accounts in part for the durability of fossils through the long episodes of geologic time. The original skeletal substance may remain chemically unchanged, or it may be replaced by mineral matter of different composition. Pores may become filled with lime or silica brought in by underground water. Entire organisms may leave vacant molds in the sediment, later to be filled so as to form casts (fig. 1.1) which show faithfully the original form. The casts are nature's commemorative statues of its former experiments.

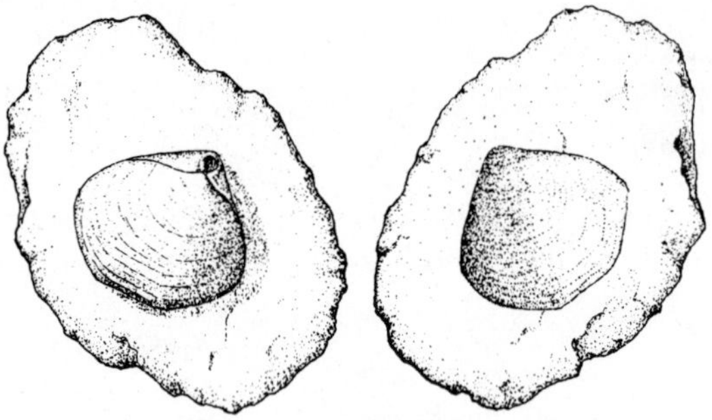

FIG. 1.1 *Cast (left) and mold (right).*

Although it is true that hard parts are more plentiful as fossils, the paleontologist is occasionally rewarded with clues to the nature of soft tissue and organs. Insects have been found preserved in the hardened resin of conifers (fig. 1.2). Carbonized remains of soft-bodied animals such as jellyfish and worms occur between the laminae of fine textured shales. X-ray examination of slabs of rock has revealed the ghostly outlines of tentacles, intestines, and even eyes of a variety of marine organisms. Soft parts such as skin, hair, and viscera of Ice Age mastodons have been preserved for us in glacial ice or pickled in the tar of oil seeps. Even frail, microscopic remains of unicellular creatures have been found embedded in the rock which hardened around their bodies hundreds of millions of years ago (fig. 1.3). Thus, the hard and soft parts, bone and shell, trails and imprints, and the many petrifactions of ancient organisms permit us to write the history of life. The story is not complete, for there are many so-called missing links, and few

FIG. 1.2 *Flower fly preserved in Baltic amber, Oligocene Age. (Courtesy of Frank M. Carpenter, Harvard University.)*

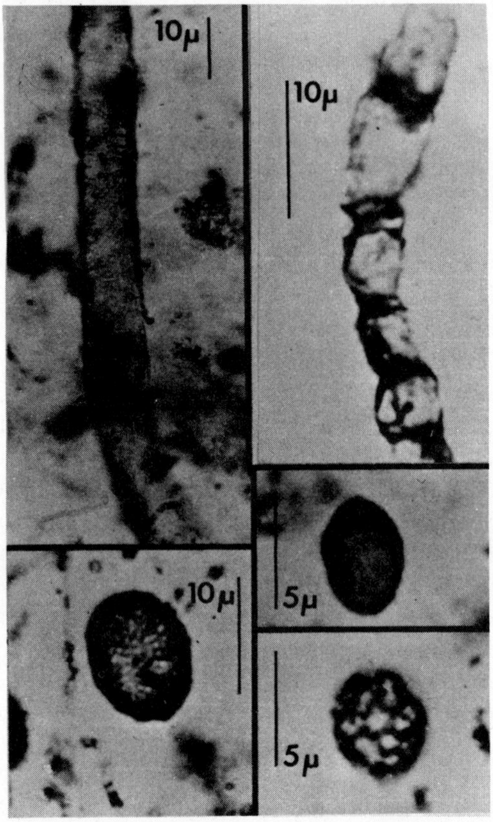

FIG. 1.3 *Microorganisms preserved in chert. From* Gunflint Chert, *Precambrian Age. (Courtesy J. Wm. Schopf.)*

of the countless individual animals of the past survive as fossils. What we have is a decidedly biased record. It is more complete for marine life with hard exterior skeletons and less complete for creatures lacking shell or bone. Usually only a fragment is left for the fossil hunter, but therein lies the fascination. The grave robber becomes the detective, using reason and imagination—and yes, intuition—to recreate the form and habit of a once living thing.

A CALENDAR

This is a book about the history of life on earth. Any history requires a calendar to which particular events or conditions can be referred. The paleontologist's calendar is called the *geologic time scale* (fig. 1.4). We are familiar with calendars used in history classes upon which boundaries are placed at times of great changes in culture and government. In geologic history, primarily there are modifications in the overall characteristics of life, and secondarily, additional modifications at times of profound changes in the earth's crust. The relative dates of these changes are determined by the superpositional relationships of fossil-bearing strata (oldest at the bottom, youngest at the top), by the stage of evolutionary advance shown by the fossilized organisms, and by absolute ages of rocks as revealed by analyses of decay of radioactive atoms.

The two greatest divisions in the geologic time scale are termed *eons.* Approximately seven-eighths of earth's history occurred in the first eon—the Cryptozoic. To the geologist, the phrase "in the beginning" alludes to this Cryptozoic interval of time which began about 4.5 billion years ago. During early Cryptozoic time, the earth had amassed most of its rocky substance, possibly from some far older turbulent cloud of cosmic dust. Some time later, but still within the Cryptozoic, the first life appeared. As we shall see in a later chapter, the evidence for this ancient life is meager. It was soft and microscopic. Most traces of our primordial ancestors were erased by destructive geologic processes occurring over billions of years.

By slow degrees, the primitive life of the Cryptozoic gave way to the more diverse, more complicated, and more abundant life of the next eon—the Phanerozoic. By studying the relatively abundant fossil record of the Phanerozoic, paleontologists are able to recognize three major subdivisions. These subdivisions are termed *eras.* The oldest era is the Paleozoic which lasted about 370 million years. The Mesozoic Era followed the Paleozoic and lasted 170 million years. The Cenozoic Era, in which we are now living, began about 60 million years ago.

As shown in figure 1.4, the eras have, in turn, been divided into shorter time intervals called *periods.* The rocks deposited during a period are known as a *system.* Each of the geologic systems is recognized largely on the basis of the fossilized animals and plants that lived during that specific time. For example, the Tertiary System characteristically contains large numbers of modern-looking marine invertebrates, whereas the marine organisms of the Cambrian were considerably different from those living today. Each system was characterized by its own fauna and flora, different in the stage of evolution (and hence appearance) from both older and younger systems.

Time Units of the Geologic Time Scale (Numbers are absolute dates in millions of years before present, after Faul, 1966)					Distinctive Records of Life (From Longwell and Flint, *Introduction to Physical Geology:* John Wiley and Sons, 1962)
Eras	**Periods**		**Epochs**		
CENOZOIC	Quaternary		(Recent) Pleistocene	2.5	Development of man
	Tertiary		Pliocene	7	Dominance of elephants, horses, large carnivores
			Miocene	26	Development of whales, bats, monkeys, horses
			Oligocene	40	Grazing animals widespread, grasses abundant
			Eocene		Primitive horses
		65	Paleocene	70	Rapid development of mammals
MESOZOIC	Cretaceous	135			Extinction of dinosaurs and ammonites. Development of flowering plants
	Jurassic	190			Climax of dinosaurs Earliest record of birds
	Triassic	225			First primitive mammals; conifers and cycads abundant; appearance of dinosaurs; rapid development in reptiles
PALEOZOIC	Permian	270			Development of conifers Spread of reptiles
	Pennsylvanian (Upper Carboniferous)	310			Earliest primitive reptiles Abundant insects Coal-forming forests widespread
	Mississippian (Lower Carboniferous)	350			Echinoderms abundant Spreading of fish faunas
	Devonian	400			Appearance of amphibians Earliest forests
	Silurian	430			Earliest record of land plants and animals
	Ordovician	500			Primitive fishes, the first known vertebrates
	Cambrian	600			Marine invertebrate faunas abundant; trilobites predominant
PRECAMBRIAN TIME					Wide distribution of simple marine plants

Relative Lengths of Major Time Divisions in True Scale

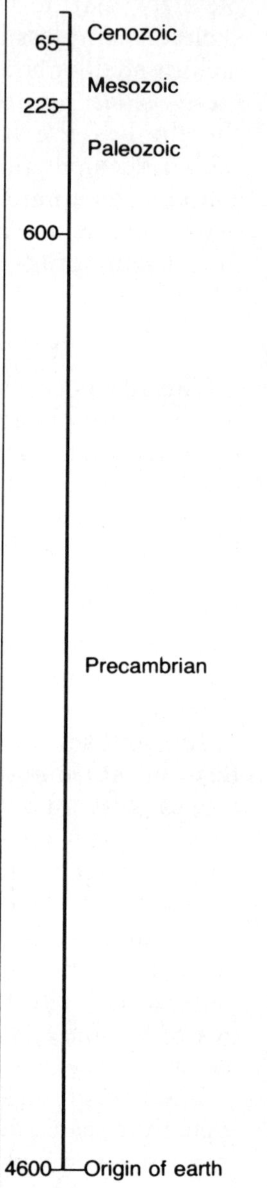

FIG. 1.4 *Geologic Time Scale. (Adapted from J. C. Brice and H. L. Levin, 1969,* Laboratory Studies in Earth History, *Wm. C. Brown Company Publishers, p.* iv.*)*

Thus, before men knew how to date rocks containing radioactive minerals, they took advantage of this constant change in living things to order geologic history. Rainstorms and floods, earthquakes and volcanoes have come without novelty from age to age. Mountains have been crushed upward, worn down to their very roots, and then uplifted again while rivers spread their fragmented essence to the lowlands and to the sea. Those sediments—those rocks—did not differ markedly through time. But life continuously presented new faces as older ones disappeared. Species did not live again in the same guise in later ages as did the lavas and sands of a restless earth. Fossils permanently haunt the rocks of their particular age and act as signposts for the traveler through time.

THE RANK AND ORDER OF LIFE

Because of the large numbers of organisms that have lived or are living now, random naming would have resulted in much confusion and seriously impeded scientific progress. Realizing this, the Swedish naturalist Carolus Linnaeus (1707–1778) established a system for naming animals and plants. The Linnaean system recognized structure as the basis of classification and utilizes binomial nomenclature at the species level. Thus, the name of a species consists of two parts, the generic name and the trivial (species) name. For example, the scientific name for a dog is *Canis familiaris.*

The basic category in this hierarchical system is the species. Very simply, species may be defined as a population of similar individuals, alike in their structure and functional characteristics, and which have a common ancestry. The individuals, which are the ultimate items of classification, can thus be grouped into species, the species combined into genera, the genera into families and so on. To use an example familiar to all, man is a member of the Phylum Chordata, Subphylum Vertebrata, Class Mammalia, Subclass Eutheria, Order Primates, Family Hominidae, genus *Homo,* species *sapiens.*

It is important to understand that biological classification is much more than a cataloging system. For the paleontologist, the basis of classification is structure—the size, form, and arrangement of parts of an organism. Structures tell the story of ancestral evolutionary relationships, and thus permit us to reconstruct the family tree of life.

SUMMARY

Paleontology is the science of life that existed in the geologic past. It involves the study of the remains and traces of ancient animals and plants. Such former indications of life are called fossils. Fossils tell us about the distribution of seas, shores, and mountains hundreds of millions of years ago. They tell us of changes in climate, depth of water, land and sea barriers, and former locations of continents. They provide evidence for the concept of organic evolution.

Fossils are preserved by a variety of processes such as chemical replacement and filling of pore spaces with mineral matter. Many fossils are impressions or molds made in once soft sediment, or they may be the casts made by the filling of previously existing molds. Whatever the process, animals with hard parts of teeth, bone, or shell are more commonly preserved than are creatures composed of easily decayed soft tissue.

Many fossils are an index to the age of the rock in which they are found. They are a record of life which inhabited the earth at the time the enclosing sediment was being deposited. Geologists have observed that there is a regular succession of fossil floras and faunas the world over. This observation, coupled with the knowledge that in any pile of undisturbed strata the oldest bed is at the bottom, was the original basis for describing the history of the earth in terms of a relative geologic time scale. The basic unit of the time scale is the geologic period. Subdivisions of periods are termed *epochs.* Periods may be lumped into larger units called eras, and eras combined into eons. The geologic time scale is a relative scale in that each unit is "relatively" younger or older than units above or below. The determination of the actual age in years requires the use of radioactive isotope dating methods. Such methods have provided ages in millions of years to the divisions of the geologic time scale.

Confronted with the great diversity of life, men have attempted to classify organisms into groups and relate the groups to one another. The system used for over two hundred years was originally proposed by Linnaeus in 1758. It is a binomial system in which the scientific name identifying an organism has two parts: the name of the genus (generic name), and the name of the species (specific or trivial name). The species is the basic category in a hierarchy of terms rising in rank from species through genus, family, order, class, phylum, and kingdom.

SUPPLEMENTARY READINGS AND REFERENCES

Beerbower, James R. 1968. *Search for the past.* 2nd ed. Englewood Cliffs, N.J.: Prentice-Hall.

Dunbar, C. O., and Waage, K. M. 1969. *Historical geology.* 3rd ed. New York: Wiley.

Easton, W. H. 1960. *Invertebrate paleontology.* New York: Harper and Bros.

Eicher, Don L. 1968. *Geologic time.* Englewood Cliffs, N.J.: Prentice-Hall.

Fenton, C. L., and Fenton, M. A. 1958. *The fossil book.* Garden City, N.Y.: Doubleday.

Raup, D. M., and Stanley, S. M. 1971. *Principles of paleontology.* San Francisco, Calif.: W. H. Freeman.

Seilacher, Adolf. 1964. Biogenic sedimentary structures. In *Approaches to paleoecology,* ed. J. Imbrie and N. Newell, pp. 296–313. New York: Wiley.

Thenius, Erich. 1973. *Fossils and the life of the past.* Heidelberg: Springer-Verlag.

2 Seeds of Change

There is nothing constant in the universe,
all ebb and flow, and every shape that's born
bears in its womb the seeds of change.
—Ovid, *Metamorphoses,* XV

" WE MAY BE ALL MELTED TOGETHER."
—*Charles Darwin*

There is an expression which asserts with the power of dogma that "nothing is fixed in nature." The phrase is an apt one, considering the lengthy record of life as it is revealed by fossils. One cannot help but notice that the lower orders of living things appeared first, and that progressively higher and higher types developed in succession. At first there was only the chemical and physical potential for life. Next, after profound changes, primitive single-celled organisms arose, perhaps from tiny spheres which today would defy classification as either living or nonliving. From these microspheres, more advanced plants and animals evolved. For a time, invertebrate animals were the only representatives of the animal kingdom. Then fishes appeared and, for a while, were the highest forms of vertebrate animals in existence. In even younger ages, amphibians, reptiles, birds, and mammals joined in the parade of advancing life. If we examine the family tree of any group within these larger categories, we see evidence of constant changes and increasing diversity through time. Yet throughout these transmutations runs a thread of unity, for all living things are descendents of forms that lived in the past. No matter how complex they have become, they all exhibit basic similarities in their solutions to fundamental life processes.

The doctrine of evolution, as it has sometimes been called, is based on a body of scientific observations indicating that the organisms of today are descendents of those that lived in the past; and that the whole complex, modern, organic world developed gradually from simpler and older forms of life. In a sense, evolution is an expression of nature's ability to change organisms over many generations and thereby make them adaptable to changing environments.

Fossils provide us with evidence of evolutionary change. Modifications in organisms become apparent in studying successively younger strata of sedimentary rocks. It is the task of the paleontologist to infer the reasons and causes for those modifications. Therefore, it is useful to review a few of the concepts that lead us to better understand these underlying causes of evolution.

THE EARLY EVOLUTIONISTS

The concept of progressive modification of animals and plants is not new. Interest in evolution can be traced back to the time of Aristotle (384–322 B.C.). Later, in the Western world complete faith in the scriptural account of the Creation stifled thinking about evolution. However, the concept of special creation was ultimately to be challenged. The French zoologist Buffon (1707–1788) and the Swedish botanist Linnaeus (1707–1778) were among the first to suggest some sort of evolution among organisms subsequent to the creation. About half a century later, the gifted and articulate Jean Baptiste de Lamarck (1744–1829) set down the first general theory of evolution. He was followed a half century later by Charles R. Darwin (1809–1882) and Alfred R. Wallace (1823–1913). Lamarck's theory of evolution, submitted in final form in 1809, embodied the following two ideas: (1) New structures appear because of "inner want" of the organism, and (2) Those structures acquired in response to need are then inherited by later generations. Almost from the first, Lamarck's ideas were challenged. His first point was scientifically untestable, and the second, when tested, was proven false. We know that animals can "adapt" in a narrowly limited way to their environment within a finite life span. The white-skinned coed on the beach can be protectively made tan. However, no amount of sunbathing will result in her first child being born with a suntan. In short, there is no known way that somatic or body cells can pass characteristics to reproductive cells. More precise experiments show clearly that there can be no inheritance of acquired characteristics. In addition, many changes found in organisms are not of a type that could be acquired, even if Lamarck's use or disuse concept was valid. Thus, the Lamarckian theory of evolution was largely discredited. Nevertheless, this great French naturalist merits our applause. He perceived that lineages of organisms do change, that more complex creatures may arise from simpler forms, and that the changes require many generations and great lengths of time. Unfortunately, he failed to provide a suitable theory to explain how evolution occurred.

Charles Darwin (1809–1882), building on a framework of ideas scattered among the thoughts of Lamarck, Lyell, and his grandfather Erasmus Darwin, provided another step toward our present understanding of organic evolution. Alfred R. Wallace, also a naturalist, shared in the revelation. Darwin's great work entitled, *The Origin of Species by Means of Natural Selection,* appeared in 1859. It came at a time when the European intellectual atmosphere was more liberal, more dissatisfied with "special creation," and more predisposed toward consideration of a new materialistic theory of evolution. As a result, the book had a profound influence upon men's thinking and deserves its place among the masterpieces of all time.

The biologically important aspects of the Darwinian theory are embodied in the following factual observations. (1) The potential for reproduction of any species far exceeds that necessary to merely maintain its numbers. Hence, more plants and animals are produced than can survive. (2) Variations of all grades exist in these plants and animals. (3) Competition for food and living space results inevitably in the elimination of the weaker and less well-adapted individuals with simultaneous "preservation of the favoured races."

Darwin deserves our acclaim for more realistically accounting for those gradual changes in the characteristics of a population that might ultimately result in its transformation into different species. His book contains a great mass of highly persuasive evidence and numerous examples collected during his five years as naturalist aboard *H.M.S. Beagle.* "Special creation" no longer seemed a tenable hypothesis among educated Victorians and *The Theory of Natural Selection* rapidly began to accumulate converts. That some of Darwin's contemporaries should have considered his concepts sinister seems unwarranted. For Darwin, there was "grandeur in this view of life, with its several powers, having been originally breathed by the Creator into a few forms or into one; and that, whilst this planet has gone cycling on according to the fixed law of gravity, from so simple a beginning endless forms most beautiful and most wonderful have been and are being evolved."

Most theories have a weakness, and Darwin's was no exception. His theory failed to account for the cause of variability upon which natural selection acts. Darwin realized this and struggled to find the key to the problem. The key lay in an understanding of heredity. It is now legend that the key was held in the mind of a Moravian monk named J. Gregor Mendel (1822–1884). Mendel discovered the basic principles of inheritance and published his findings in an obscure 1865 publication. The work was unknown to Darwin and was ignored by the scientific community until its rediscovery about 1900. Today, every knowledgeable person has heard about Mendel's breeding experiments with garden peas. Mendel demonstrated that the heredity of the peas was controlled by paired determinants which divide in the pollen and ovules and are recombined within certain finite limits in the course of fertilization. Stimulated by this discovery, knowledge of heredity increased rapidly. During the first third of the twentieth century, satisfactory hypotheses for the origin and persistence of variability began to emerge.

In our thinking of evolutionary concepts, there is an important point to bear in mind. It has to do with an understanding of the meaning of the term adaptation. Adaptation is a word used by biologists to indicate any characteristic of an organism or species that is of value to its survival or reproductive success. The word is also applied to the process by which beneficial characteristics are achieved. The term is not intended to imply a Lamarckian mechanism for evolution. When one speaks of certain creatures or species "adapting" to an environment, one is simply employing a useful conversational expedient which decidedly does not suggest that the creatures involved made a conscious or deliberate effort to alter themselves so as to better cope with the demands of their surroundings. Rather, adaptations are the result of complex modifications of hereditary determinants in the population through the effects of survival and breeding. These effects are in turn related to environmental influences, and the outcome of unpredictable and random events.

DETERMINANTS OF CHANGE

The work of Mendel and his followers over the past seventy years has shown that the development of the individual organism is controlled by heredity determinants or regulators called *genes.* It has been demonstrated that genes are composed of deoxyribonucleic acid (DNA). A single gene directs the formation of a specific enzyme which in turn passes out of the nucleus into the surrounding portion of the cell and serves to regulate some specific factor in the cell's metabolism, growth, or reproduction. The genes are also capable of exact self-replication.

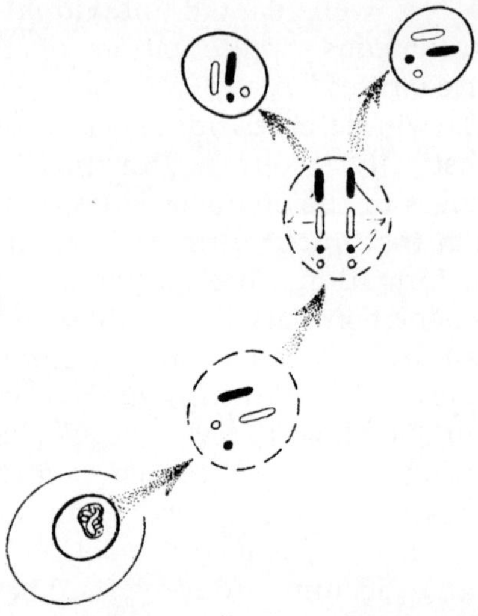

FIG. 2.1 *Schematic summary of mitosis. (Adapted from J. M. Savage, 1969,* Evolution, *Holt, Rinehart, & Winston, Inc., p. 19)*

Genes in most organisms are arranged in line along paired filaments called *chromosomes.* The type and number of chromosomes are constant for each living species and differ between species. Except in the blue-green algae and bacteria, chromosomes are located in the nucleus of the cells. When counted and marked against one another, it is usually found that chromosomes occur in duplicate pairs. That is to say, each individual chromosome possesses a homologous mate with an identical size and shape. Thus, man has forty-six chromosomes which actually represent twenty-three pairs. Cells (and organisms) with paired homologous chromosomes are spoken of as being *diploid.* In all living things, new cells are being produced constantly to replace worn out or injured cells and to provide for growth. In asexual organisms and in the cells of the body other than sex cells, the process of cell division that produces new diploid cells or individuals with exact replicas of the chromosomal and genetic components of the parent cells is called *mitosis* (fig. 2.1). The process can be viewed under a microscope. Concentrating on the nucleus, one initially observes swarms of gray granules which begin to assume arrangements as slender threads of varying lengths. These, of course, are the chromosomes. Soon the chromosomes become short and stubby and arrange themselves near the middle of the cell. Each chromosome splits in half. Meanwhile, threads form in the fluid (cytoplasm) and gradually pull the halves apart until the chromosomes are separated into two groups at either end of the cell. A membrane develops across the cell, dividing it into two daughter cells. The chromosomes in each half gather themselves into a nucleus, each with an exact replica of the chromosomal and genetic components of the parent nucleus.

There is another type of division called *meiosis* that takes place when gametes (egg cells or sperm cells) are formed (fig. 2.2). Meosis may occur in unicellular organisms, but in multicellular forms it takes place only in the reproductive organs. In man, for example, cells are produced by mitosis throughout the body except in the gonads (testes or ovaries). Meiosis involves two cell divisions. In the initial division, the chromosomal pairs are divided in random fashion so that each of the two new cells receives

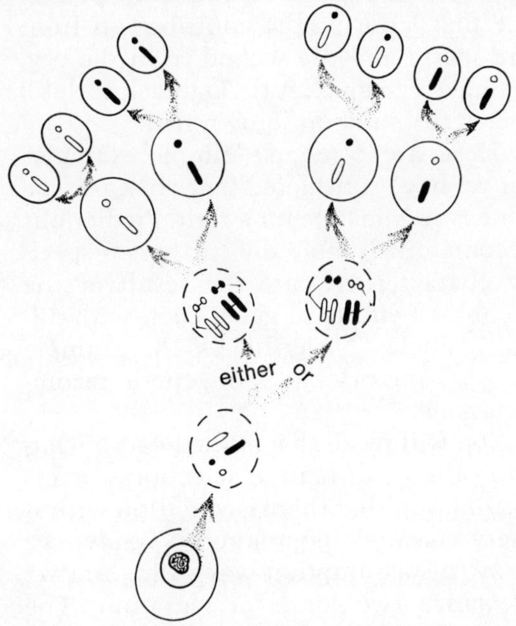

FIG. 2.2 *Schematic summary of meiosis. (Adapted from J. M. Savage, 1969,* Evolution, *Holt, Rinehart, & Winston, Inc., p. 20.)*

one member of each original pair, or precisely half the number of chromosomes found in the original cell. These daughter cells are termed *haploid* because they do not have paired chromosomes. The haploid cells reproduce themselves exactly as in mitosis to make a total of four cells. These four cells are the haploid gametes or reproductive cells. Thus in humans, the egg and sperm cells (gametes) contain not forty-six chromosomes but twenty-three. Each of these gametes contains half of the number of chromosomes in the original diploid parent cell. When two gametes meet during sexual reproduction, the sperm enters the egg to form a single cell which can now be called the fertilized egg. Since two haploid gametes have now fused into one, there are now a full complement of chromosomes. By repeated mitotic divisions the fertilized egg initiates the development of a complete organism.

There is another aspect of meiosis that is important in producing variation in offspring. During the initial division while the doubled chromosomes are still paired, they seem to break at corresponding places and exchange their severed parts in a process called *crossing over.* The result, of course, is still further mixing of the genes. They are segregated and recombined in random fashion so as to produce endless variety of combinations of genes generation after generation. The process can be compared, within limits, to the shuffling and dealing of playing cards. One pack of fifty-two cards can yield a variety of hands. Just as in poker where a full house beats three of a kind, so it is that in this evolutionary poker, certain combinations of genes may fortuitously turn out to be more beneficial than others and are favored by natural selection.

It is now apparent that the appearance of an individual is ultimately determined by the genes. We refer to the appearance of an individual as its *phenotype*—the visible finished product. Mendel's peas, for example, produced either "round" or "wrinkled" pea phenotypes. The expression or enzyme function of the phenotype in the gene is called the *allele* (pronounced uh-*leel*). In an organism, any combination of alleles may be present. Mendel's peas, for example, may have had two round pea alleles or two wrinkled pea alleles. This does not mean that the alleles are round or wrinkled, but only that round allele controls the formation of an enzyme involved in one of the steps in the production of round peas in the pod. For illustration, we can call the round pea allele *r* and the wrinkled pea allele *w*. The possible combinations are thus *rr*, *ww*, or *rw*. The actual combination of the genetic materials is called the *genotype*. It is apparent that the *rr* genotype would produce a rounded pea (in the pod) phenotype and the *ww* genotype would produce a wrinkled pea phenotype. But what of *rw*? In this case, either kind of phenotype might result. If the plant developed with rounded peas in its pods, then the wrinkled allele was masked and is thus termed *recessive*. The rounded allele may then be called *dominant*. In the formation of reproductive cells, individuals of genotypes *rr* and *ww* each produce only one kind of gamete because of the separation which occurs during meiosis. The gametes will be all *r* or all *w*. A plant or animal producing only one kind of gamete is called *homozygous*. The organism with a genotype like *rw* will produce either *r* or *w* gametes and thus is called heterozygous.

Let us consider a few of the possible combinations using mating pairs of guinea pigs. We begin with the mating of two homozygous individuals, a "pure" brown (*bb* phenotype) male with a "pure" black female (*BB*). During meiosis in the male, the two *bb* genes separate so that each sperm has only one *b* gene. Similarly the dominant *BB* genes in the female separate so that each ovum has only a single *B* gene. The fertilization of this egg by a *b* sperm results in an animal with the genetic formula *Bb* and black fur. Now let us consider a heterozygous pair of guinea pigs. During meiosis in heterozygous animals, the *Bb* chromosome separates so that each egg or sperm has either a *B* gene or *b* gene but never both. Since there are two types of eggs and two types of sperm, four combinations are possible in fertilization. Letting the first letter be the contribution from the sperm and the second from the egg, we have *BB* (black fur), *Bb* (black fur), *bB* (black fur), and *bb* (brown fur).

These are, of course, simple examples for we have considered that only a single gene is responsible for a trait. Predictions become much more difficult when specific characteristics are the result of the combined effects of genes each with differing alleles. Nevertheless, the example suggests the principles of genetic recombinations.

You will recall that at the heart of Darwin's theory of natural selection was his assumption that there is variation within every natural population. Today we know his assumption was valid, and we recognize two kinds of variation. The first of these is the variation acquired during the lifetime of an individual. The muscles of the weight lifter or suntan of the sunbather are examples of variations caused by environmental influences. As we have seen, such variations are not important in evolution. What are important are variations resulting from hereditary differences, for only such differences can be passed along to succeeding generations. Hereditary differences are the raw materials of evolution. We have seen in a simplified way how gene recombinations and crossovers cause variation. But one cannot help wonder how genetic variation came about in the first place, or how a completely new trait appears in a population. Mutations provide us with an answer to these questions. Simply stated, mutations are sudden changes in the genetic material within cells. The altered or mutant gene or portion of a chromosome produces an effect different from that possible with the original genetic elements.

Mutations may occur at one or more of the organizational levels within the genetic material. At the lowest level, it may alter the building block sequence of DNA in one or more genes. At a higher level, the mutation may result from a structural alteration of the chromosome in which the DNA is arranged. At a still higher level, a mutation may result from an increase in the number of chromosomes. Mutations can occur in any cell, but their evolutionary impact is greatest when they occur in the gametes, for then they will affect the succeeding generations. Mutations can be caused by external factors like ultraviolet light, certain foods and drugs, X rays, and various other forms of radiation or chemical substances. Moreover, some mutations may occur spontaneously in the absence of a specific causative agent.

Mutations are the ultimate source of new and different genetic material appearing in a population and thus can produce evolutionary change. Recombinations of genes enhance the effect of mutation by producing a broad spectrum of possible gene combinations. Recombination alone could not produce evolution as we know it, since the possibilities within an unchanging collection of genes are limited and finite. Nevertheless, by distributing the effects of mutation for natural selection to act upon, it cannot be denied that recombinations are also important to evolutionary change. Clearly organic evolution requires a triumvirate involving gene recombinations, mutations, and natural selection. The latter sorts the multitude of traits, preserving those opportunistically best fitted to a particular environment.

EVOLUTION IN POPULATIONS

Struggle and survival were important in the Darwinian concept of natural selection, but it is now recognized that neither the actual physical struggles and competitions, nor the individual change in a few characteristics from parent to offspring are as important as Darwin and Wallace believed. A larger view is required. For example, when a paleontologist thinks about evolution, he conjures up images like that of the family tree of camels or horses. In his mind's eye, he sees certain overall trends in size and morphology that provide a framework for the more numerous variations that exist among the various species of camels. He takes the broad view provided by the perspective of time and tends not to concern himself with the difficult-to-detect individual variations that may occur between closely spaced generations. The accumulative changes in the entire population of camels or horses are the basis for study. To the paleontologist, sustained evolution is true evolution, and sustained evolution is accomplished only by the accumulation of mutations that expand and diffuse—thereby altering entire populations. The basic unit of evolution is the population.

This new aspect of organic evolution has become known as population genetics. In population genetics, an attempt is made to describe the genetic makeup of an entire population. The population is considered to be composed of all the individuals of the same species occurring in the same area at a particular time. The genetic characteristics of that population can then be treated as the sum of the genotypes of all the individual members. The sum of these genotypes constitutes the *gene pool* of the population. The number and kind of each allele found in the population determine the unique characteristics of the gene pool. An important concept in population genetics is that individual variations in organisms are not considered unless they become so prevalent as to be statistically significant. Only if enough organisms with the same variation coexist in a given generation will this variation become apparent as a definite and nonrandom change in the gene pool's composition.

By exerting influence on an individual organism, natural selection ultimately brings about a change in composition of a gene pool to which these individual organisms contribute. The gene pool of a population is divided up each generation and partitioned out to new offspring. Inevitably, in each new generation, new mutations or combinations of genotypes occur which express themselves as specific individual phenotypes. As a result of natural selection certain genotypes will leave more offspring than others. Certain genes will be passed on to the next generation in greater numbers, others in lesser, so that the composition of the gene pool will be changed accordingly. The fundamental source of evolutionary change becomes the impact of natural selection upon the gene pool.

Two antagonistic factors operate on the gene pool. The first of these is natural selection which tends to alter the composition of the gene pool from one generation to the next. The other factor is *genetic equilibrium.* Genetic equilibrium asserts that if there is equal opportunity of producing offspring in a large, randomly mating population, the ratio between various alleles in a population tends to remain constant from generation to generation. The concept was brought to light independently by G. H. Hardy and W. Weinberg, and has become known as the Hardy-Weinberg Law. Genetic equilibrium expresses the stability and conservatism of geological heredity. It favors the status quo. Hence, evolution cannot occur until the equilibrium is upset. It may be upset if conditions change within a population through the origin of additional variability or through external environmental changes. Fundamental forces of evolution like mutation and natural selection may disrupt equilibrium and bring about evolutionary change.

ADAPTATION—EXPAND AND CONQUER

Adaptation can be simply defined as the process by which an organism becomes better fit for its environment. An adaptation can be considered a characteristic which enables the organism to survive or excel in a particular environment. Adaptations develop according to the mechanisms discussed above whenever successive generations of a particular lineage move into slightly different environments, or when a particular geographic habitat undergoes changes. Certain organisms are thereby provided with the opportunity to spread into a number of new ecological sites to which they have physical access, and in which there is a possibility of survival. In these new sites, over many generations, adaptations will occur. Because of the competition for food and living space, there is a tendency to spread out and occupy as many different habitats as can be reached, and in which survival is assured. As more and more specializations and adaptations arise from the less specialized ancestral species, forms which are morphologically and physiologically adjusted to the diverse sites become distinctly recognizable. The process is called adaptive radiation. The essence of adaptive radiation is the evolution of a variety of different forms from a single ancestral form, with each of the descendent lines being adaptively specialized in a unique way to survive in a particular habitat.

The history of the Class Reptilia provides a classic illustration of adaptive radiation. The earliest known reptiles lived in Pennsylvanian times. They represented a shift from partially aquatic to a completely terrestrial habitat made possible by the development of the amniotic egg. These early reptiles, called cotylosaurs, provided the ancestral stock for the adaptive radiation of the marvelously diverse array of Mesozoic reptiles. The cotylosaurs were themselves the descendents of Paleozoic amphibians.

It appears likely that the breakthrough that produced the earliest reptiles—that moved vertebrate evolution into a new adaptive zone—followed thousands of years of evolutionary experimentation among amphibians. One or more of these amphibian "experimental lineages" managed to break its ecological barrier and enter the new territory. It is likely that the amniotic egg, ordinarily the accomplishment of reptiles, evolved among certain species of amphibians, thereby providing greater safety for fertilized eggs.

The history of life tells us that such breakthroughs are usually geologically rapid. In the case of the amphibian-reptilian transition, it appears to have taken place in a fraction of a geologic period. Rapid diversification followed the breakthrough. Such is usually the case as new zones, initially deficient in competition, are opened for biological exploitation. As the vertebrates moved from their formerly aquatic habitat to the land, new adaptations appeared such as dessication-resistant skins and gravity opposing skeletal and muscular modifications. After these critical changes were accomplished, the reptiles continued their radiation, producing flying reptiles, marine reptiles, and a variety of specialized land-dwelling forms. Among the latter, another major barrier was eventually penetrated as the first bird evolved from a lightly built, bipedal reptile. Thus, we see that evolution is constantly in motion, ever changing, always probing at its boundaries, experiencing vast divergences, and continuously adjusting to the enormously complex interaction of physical and biological forces affecting the surface of the earth.

SUMMARY

Simply stated, organic evolution is the concept that all living organisms are the descendents, with modifications, of life that lived in the past. Because evolution involves change over long periods of time, those changes should be, and indeed are, represented in the chronological succession of fossil organisms. Among the founders of our modern concepts of evolution are Darwin and Wallace, who proposed the theory of natural selection; Mendel, who provided an understanding of the laws of heredity; and De Vries, who developed the mutation theory of evolution.

The development of an individual organism and the characteristics of that organism are controlled by chromosomes and genes. The chromosomes are located in the nuclei of cells, carry the genes that are composed of DNA, and are responsible for the development of characteristics that make an animal or plant alike or different from others. Meiosis segregates the members of homologous pairs of chromosomes and halves the total number for each gamete. During fertilization, there is a random union of two gametes from parents of unlike sex which brings together assortments of chromosomes (and genes) from two parents, resulting in the production of individuals with different gene combinations. Changes in genes called mutations may occur. Changes may occur also by rearrangement of chromosomes. These changes and redistributions alter the assortment of genes and, hence, also the characteristics which are passed on to succeeding generations.

Population genetics involves the application of Mendelian principles to entire natural populations. The Hardy-Weinberg Law is important to an understanding of population genetics. In brief, this law states that in a given population with random mating and no selection between genotypes, the relative frequency of alleles (contrasting genes in homologous chromosomes) tends to remain constant from generation to generation. Under such conditions there can be no evolution. Evolution is the deviation from this norm, caused by selective mating, mutations, and the survival of individuals with useful characteristics that result in change.

Adaptation is the process by which organisms become fitted to their environments. Commonly, adaptations involve a combination of characteristics, including not only structure, but physiology and behavior as well. Different vertebrates, for example, display adaptive modifications of limbs for either swimming or running, or of teeth for various kinds of food. The evolution and spread of a single lineage of organisms into several ecologic niches, often results in different forms, each of which is distinctly adapted to a particular habitat. This process of diversification from a more or less central stock is called adaptive radiation.

SUPPLEMENTARY READINGS AND REFERENCES

Darwin, Charles. 1859. *The origin of species.* 1963 ed. Intro. by Hampton L. Carson. New York: Washington Square.

DeBeer, G. 1964. *Atlas of evolution.* London: Thomas Nelson and Sons.

Dobzhansky, T. 1950. The genetic basis of evolution. *Scientific American* 182:32–41.

———. 1955. *Evolution, genetics, and man.* New York: Wiley.

Eiseley, Loren. 1961. *Darwin's century.* Garden City, N.Y.: Anchor Books, Doubleday.

Levine, R. P. 1962. *Genetics.* New York: Holt, Rinehart & Winston.

McAlester, A. L. 1968. *The history of life.* Englewood Cliffs, N.J.: Prentice-Hall.

Savage, Jay M. 1963. *Evolution.* New York: Holt, Rinehart & Winston.

Volpe, E. P. 1967. *Understanding evolution.* Dubuque, Iowa: Wm. C. Brown.

3 The Primordial Earth

Then was not non-existent nor existent: there was no
realm of air, no sky beyond it.
What covered in, and where? And what gave shelter?
Was water there, unfathomed depth of water?
—"Creation," from *Rig-Veda,* Hindu Scripture

DUST CLOUD TO PLANET

Before there was life, there had to develop a suitable planetary environment for life. There had to have been an earth. Theories on the origin of our planet differ in details, but most agree that the earth and other planets in our solar system formed from a turbulent cloud of swirling cosmic dust that surrounded our sun star. The dust originally accumulated in the voids of space, driven to relatively empty regions by the radiant energy of stars. The cloud probably consisted of myriads of elemental particles; gases such as hydrogen, helium, nitrogen, and neon; and simple, often frozen, gaseous compounds (carbon dioxide, methane, ammonia). Astronomers tell us that the dark areas seen silhouetted against the more luminous gaseous nebulae are similar clouds of cosmic dust. The dust cloud hypotheses for the origin of the solar system would begin with such a region of nebulous, finely dispersed matter. Gradually particles within the cloud began to gravitate inward so as to form a smaller denser body. A slow rotation began. The stimulus for this rotation may have been derived from rotational movement of the parent galaxy, radiation from nearby stellar bodies, unbalanced gravitational forces, or the movement of the particles themselves as they were drawn together. As material continued to gravitate inward, the cloud's circumference diminished; and, in accordance with the laws governing the conservation of angular momentum, it rotated ever faster. Gradually, the form of the cloud changed from roughly spherical, to oblate, to the discoidal shape now approximated by the orbital planes of the planets in our solar system. In accordance with Kepler's third law, the period of revolution of the orbiting particles would vary with the distance from the center of the dust cloud. The more central region of the cloud turned much faster than the outer. As a result, the cloud tended to differentiate into concentric zones or shells. As each of these zones revolved at different velocities, friction occurred at their borders. The friction in turn is assumed to have created turbulent eddies, smaller scuds of dust within the larger cloud.

The astronomer C. F. von Weizsäcker postulated the characteristics of such eddies over thirty years ago and prepared a mathematical model showing their size, location, and direction of movement. More recently, his theory has been modified and expanded by Gerald Kuiper, and the revised theory is now known as the *protoplanet hypothesis*. The theory suggests that on such a spinning disc, some eddies would rotate in retrograde directions. These retrograde eddies would in turn generate prograde eddies or countercurrents corresponding to the prograde rotational directions of the planets in our solar system. The constituents of the prograde eddies could then be collected into larger and larger particles. Like great whirlpools of fog, they may have swept up the smaller bodies in their paths, until, at last, aggregates somewhat larger than the proportions of planets had been accumulated. These were the protoplanets. One of them was to become the earth.

Another consequence of the gravitational shrinkage of the cloud was a general increase in temperatures in the region now occupied by the sun. Ultimately the heat became so intense (several million degrees Celsius) that nuclear fusion began. Nuclear fusion or synthesis is a reaction that occurs at very high temperatures during which helium may be built up from hydrogen. This conversion of hydrogen into helium is regarded as the basic energy-producing reaction in the sun. In this early stage of its evolutionary development, the sun began to pour forth a steady stream of radiant energy. Solar radiation, in turn, ionized the remaining parts of the dust cloud, and these now-ionized particles interacted with the sun's magnetic field, slowing its spin to approximately its present speed. The sun thereby transferred much of its original angular momentum. In addition, the angular velocities of outwardly radiating particles redistributed a portion of the solar system's angular momentum to the outer planets where we now find it. The sun's energy heated the more centrally located protoplanets, forced their gases outward toward the periphery of the system, and vaporized much of the ammonia and methane "snow." Thus, today we find relatively larger accumulations of light gases and solids in the planets located farthest from the sun, and these same planets have lower specific gravities.

In summary, we find that the more widely accepted theories for the origin of the solar system suggest that the earth originated in the cold state by accumulation of the components of cosmic dust. Our earth progressed from turbulent eddy to a solid agglutinated mass of frozen gases (including hydrogen, helium, methane, and ammonia), elemental metallic particles, and stony fragments. As this mass circled in orbit, it swept up matter that had not yet been collected into larger bodies. In its early stages, therefore, the earth was an unstratified mixture. Today, geologists know that the internal structure and composition of the earth is not at all uniform throughout.

THE CRUST AND WHAT LIES BENEATH

From studies of earthquake waves, geologists have discovered that the interior of the earth is composed of concentric shells which differ in physical properties and composition (fig. 3.1). There is, for example, a core of iron and some nickel. The outer portion of this core is believed to be molten. The core is about 3,400 kilometers in diameter and is surrounded by a heavy mantle of iron and magnesium rich silicates. Sheathing the mantle is a thin outer crust of light silicates. This internal stratification of the earth into layers of various densities indicates that the originally cool, unsorted, undifferentiated globe must have been heated, perhaps to several thousand degrees Celsius. During this thermal episode, the iron which was formerly disseminated throughout the earth was melted and drawn by gravity toward the central core. The heat required to mobilize not only the iron, but nickel and other metals, is believed to have been generated within the planet. It was probably the result of a variety of processes. Gravitational compaction was one likely source of heat. More important was probably the thermal energy released during the disintegration of short-lived radioactive isotopes that were abundantly present at that early time. The heat thus generated facilitated melting and movement of ions throughout the primordial earth and resulted in the concentric layering.

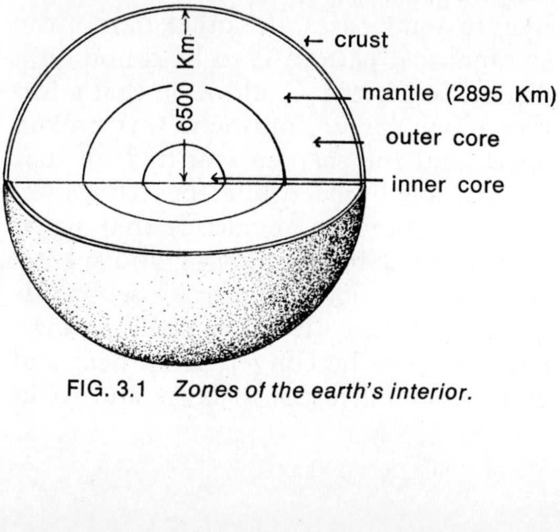

FIG. 3.1 *Zones of the earth's interior.*

The thin outer shell of this global "onion" is of special interest, for only here do living things exist. This outer shell is called the crust. Evidence which permits us to identify the lower boundary of the crust was discovered in 1909 by the Yugoslavian seismologist Anton Mohorovicic. Mohorovicic noticed that at certain depths seismographs record a distinct increase in the velocity of compressional earthquake waves. The zone in which the velocity change occurs has been named the Mohorovicic Discontinuity in honor of its discoverer. This important discontinuity separates the higher density mantle from the lower density crust.

The thickness of the crust is greater under the continents than under the oceans (fig. 3.2). The continents are com-

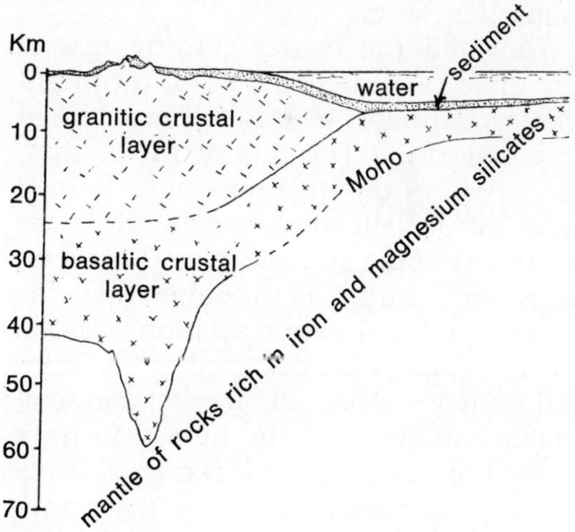

FIG. 3.2 *Diagram of the earth's crustal rocks. (After A. N. Strahler, 1971,* The Earth Sciences, *Harper & Row, p. 401.)*

posed of deep blocks of granitic rock probably resting on a basement of basaltic rock. The depth to the base of the crust under the continents is about thirty-five kilometers. Under the oceans, the granite blocks are absent, and marine sediments lie directly on the basaltic type of crust. In these oceanic regions the basaltic crust is only about five kilometers thick. The basaltic crust is heavier and contains relatively greater proportions of iron and magnesium, and relatively lesser amounts of silica than does the granitic crust.

The fact that the granitic continental crust is less dense than the basaltic oceanic crust may account for the elevations of the continents above the oceans. Somewhat like great stony icebergs, they appear to "float" in a sea of heavier basalt. The analogy can be carried a step further, for granitic crustal material beneath mountain ranges projects downward into the basalt to form mountain roots somewhat like the submerged portions of icebergs.

How did the lighter granitic masses originate? Most hypotheses for the origin of the continents tend to fall into one of two categories. The first of these stipulates that a thin crust may have formed as a sort of light silicate scum while the earth was in the molten state. This event, mentioned earlier as the time when the earth's internal differentiation was in progress, may have occurred in the interval from 4.5 to 3.5 billion years ago. Unstable conditions at the time may have caused considerable breaking, shifting about, and reassimilation of the primitive crust before it gradually acquired its present characteristics. Although the simplicity of this hypothesis is appealing, observational field work is either lacking or ambiguous. Perhaps any evidence that once existed has been destroyed by subsequent geologic forces.

The second category of hypotheses for the origin of the continents begins with a continuous basaltic crust. Here and there sections of this basaltic crust would, not unexpectedly, vary in being slightly less dense and thicker than adjacent tracts. Such anomalous regions might represent nuclei of the continents. They grew by a process called "continental accretion." The process begins with the formation of troughs along one or more sides of the continental nuclei. The trough, termed a geosyncline, receives the products of erosion and weathering from the adjacent uplands. Later, compressional forces crush the sedimentary material of the geosyncline into mountain belts. The mountain building event is accompanied by melting of portions of the wedge of sediment and by considerable metamorphism. At the end of the episode of mountain building, a marginal tract of the general composition of granite has been welded onto the continental nucleus, and the entire process can be repeated at the newly accreted periphery of the continent.

Evidence for such a concept is found in the striking architectural symmetry of North America, with its central core of ancient rocks flanked by the rocks of more recent mountain systems. On closer examination, however, the patterns are not as symmetrical as we might like. In some cases, the other half of the symmetrical pattern is to be found on a distant continent. Could it be that a former much larger continent was broken apart, and the segments drifted to their present locations? Along these lines, research is providing evidence that accretion may occur as a moving and spreading ocean floor plunges sediments downward beneath the continental margins, there to be converted by heat and compressional forces to lavas and rocks of granitic character. One aspect of this theory of seafloor spreading holds that thermal convection currents in the mantle cause it to flow like a viscous liquid and, in the process, produce geosynclines and mountains, shift continents, and transport the surface of the ocean floors.

The land animals living on these shifting continents obviously were unaware of the unstable condition of their homes, but their evolution was influenced by these shifts and fragmentations. As only one example, the more rapid and more extensive diversification of mammals as compared to reptiles may have been partly the result of reptilian evolution on large, connected land masses. Mammals, on the other hand, experienced several individual diversifications or adaptive radiations on fragmented, more or less separate, continental tracts.

Somewhat over two hundred million years ago when dinosaurs were in their ascendency, all the present continents were apparently bound into a supercontinent called Pangaea. One by one, the present continents unfastened themselves from Pangaea, riding on thick lithospheric crustal plates that were themselves driven by the slow moving convection currents mentioned previously. Some of the early breakages of Pangaea occurred in the Gulf of Mexico and between North America and Africa about 190 million years ago. After that, South America separated from one side of Africa, while Antarctica moved away from the other. As the plates and their passenger continents sluggishly moved about on the sphere of the earth's surface, new ocean floor must have been created at the lines of separation. There is ample evidence that the processes involved in plate movements are going on today, and geophysicists are able to measure the rate of movement. The lines of separation are termed "spreading centers" and are marked by high heat flow presumably from rising convection currents, as well as volcanism and earthquake activity. As new molten rock for the seafloor is extruded, it becomes magnetized in the earth's magnetic field, providing further clues of seafloor movement. This evidence is based on the discovery that the earth's magnetic field has reversed itself a number of times in the recent geologic past. The result of these reversals is that the extruded lavas form matched bands of alternate normal and reverse polarity on either side of the spreading center.

Ultimately, moving crustal plates on the spherical surface of the earth must collide. Should the line of collision involve the edges of plates which are composed of oceanic crust, a system of volcanic island arcs may be produced somewhat like the Japanese Islands. When a continental plate edge meets an oceanic one, a mountain range characterized by extensive volcanism is constructed on the edge of the continent. The Andes illustrate this type of collision. When two continental segments collide, a mountain range with few volcanoes results. Such a mountain system formed when the Indian Plate collided with Asia about fifty million years ago and produced the great Himalayan Range. Occasionally the relative motion of two plates may be parallel to their common boundary. The result is a huge lateral fault. The San Andreas Fault of California is an example. It separates the Pacific from the North American Plate.

These recent concepts of shifting

crustal plates are not without support from the science of paleontology. For many decades paleontologists have recognized that nonmarine sediments of Carboniferous and Permian age on Southern Hemisphere continents contain similar plant fossils and, in addition, much the same kind of fauna of terrestrial reptiles. Assumptions based on the skeletal characteristics of these reptiles indicates it is highly improbable that they were capable of crossing the oceans. A striking example of this type of evidence is the 1969 discovery of the remains of the mammallike reptile *Lystrosaurus* in Antarctica. These creatures have been known for years from the Karoo strata of South Africa. The presence of *Lystrosaurus* on these now widely separated land masses is certainly strong evidence for the current concepts of continental and plate drifting. These new ideas have propelled the fossil experts back to their reference collections to reevaluate former interpretations of paleogeography, paleoclimatology, and paths of migration. Fossils are providing the test information for the emerging geophysical hypotheses.

THE YEARS OF THE TEMPEST

Somewhat more than 4.5 billion years ago, the earth had accumulated most of its substance. It then entered into a period of internal differentiation and melting on a grand scale. A seared and tortured crust was blistered with volcanoes and erupting fissures with extruded molten silicates, gases, and vapors on a scale the earth never again experienced. An atmosphere compounded of these volcanic exhalations, and much different from that of today, began to accumulate. Water vapor, a major constituent of volcanic emanations, condensed and initiated the development of a hydrosphere. Overhead the sun rarely penetrated the dense turbulent cloud cover. Violent storms sent rain streaming downward, often only to be reheated and returned as steam to the atmosphere. There was no free (uncombined) oxygen, no ozone layer, and hence lethal doses of ultraviolet radiation must have been received on the surface of the earth. Hydrogen was extremely abundant. Because of the abundance of hydrogen, oxygen that did become available was immediately combined to form water. Some oxygen combined with metallic atoms to form the silicate rocks of the crust. Carbon and nitrogen were also quickly united with the ubiquitous hydrogen to form methane and ammonia. These two gases are considered to have been particularly characteristic of the earth's atmosphere during those violent early years.

Near the end of the earth's first billion years, there is evidence of a second heating of outer zones which caused their further degassing, thereby further modifying the atmosphere. The heat for this second thermal event may have been derived from radioactivity or, as geologist Preston Cloud theorizes, may have been generated by frictional forces associated with the capture of the moon. Theoretically, oxygen provided by this outgassing eliminated tremendous quantities of ammonia and methane by chemically combining with these gases. For example, oxygen can be combined with methane to form water and carbon dioxide. Similarly, oxygen can be combined with ammonia to yield water and free nitrogen. Thus a second atmosphere, rich in water vapor, carbon dioxide, carbon monoxide, and nitrogen, began to displace its methane and ammonia precursor. Still, the atmosphere lacked free oxygen, for there would have been none "left over" after the oxidation of earlier gases.

The oldest rocks providing us with evidence of this atmospheric evolution are about three billion years old. They are exposed in Africa and are clearly water laid sediments that could not have formed in the absence of an atmosphere and hydrosphere. They contain detrital particles of the minerals pyrite and uraninite. These minerals cannot form in an oxygen rich environment as they would be oxidized to other minerals. They thus suggest the absence of free oxygen. Chert, a rock composed of silicon dioxide, is abundant among these ancient African sediments. Its presence suggests the absence of ammonia in the atmosphere of 3.5 billion years ago. The alkalinity of ammonia would dissolve chert or prevent it from forming initially. Carbonate rocks such as limestone or dolomite are absent. This too indicates that the atmosphere at the time these strata were deposited was nonammoniac and somewhat acidic. In such an environment, chert could form but carbonate rocks could not. Indeed, the abundance of carbon dioxide in the atmosphere probably contributed to the formation of carbonic acids which would have dissolved carbonate minerals that may have previously formed.

There is much yet to be learned, and there is much speculation about the evolution of the atmosphere. Recent investigations, however, favor the interpretation of an ammonia and methane rich atmosphere from about 4.6 to 3.0 billion years ago. This is, of course, largely inference derived from what is known of the cosmic atmosphere. Beginning about three billion years ago, the atmosphere lost much of this methane and ammonia, and it accumulated gases derived from igneous activity and extruded from volcanoes and thermal springs. Such exhalations included water vapor, oxides of carbon and sulphur, nitrogen, and acidic vapors. Shortly we will trace the development of yet a third atmosphere, a modern atmosphere, rich in oxygen.

SUMMARY

Modern hypotheses for the origin of the earth favor the process of accretion of the planet by combining portions of clouds of dispersed gases and dusts under gravitational attraction. Once formed, the earth and other planets grew by accumulation of solid bodies of many sizes that had themselves come together in the original dust cloud. Heat, derived largely from the decay of radioactive elements, then melted rocks, caused lighter elements to rise toward the surface and heavier ones to sink toward the center of the planet. It was during this melting event that the density layering of the earth came into existence. It resulted in a concentration of iron in the core. A less dense mantle of ultrabasic rock developed above the core and was capped by a crust of still less dense rock. During this episode of differentiation, volatiles were extensively extruded, forming an atmosphere and hydrosphere. However, the early atmosphere lacked free oxygen, there was no ozone layer, and ultraviolet radiation was presumably intense. Continents began their development, possibly as the sedimentary products of weathering were remelted and solidified to form patches of lighter silicate crust. Much later, these continents and the crustal layers beneath them were to experience vast dislocations. The idea that continents may have changed their positions is embodied in the theory of continental drift. Recently this theory has been incorporated into the concept of plate tectonics. According to these new ideas, the earth's crust is divided into huge platelike segments that drift slowly upon the mantle. When two plates move apart, a fissure (called a spreading center) opens between them. Through the fissure rises hot plastic material which solidifies and joins the trailing edge of the plate. The forward edges of the plate ultimately meet another plate and are deflected downward to be destroyed by melting. Thus crustal material is being created at spreading centers and consumed at edges that plunge downward in what have been termed subduction zones.

SUPPLEMENTARY READINGS AND REFERENCES

Brandt, J. C., and Moran, S. P. 1972. *New horizons in astronomy.* San Francisco: Freeman.

Colbert, E. H. 1973. *Wandering lands and animals.* New York: E. P. Dutton.

Cloud, P. 1968. Atmospheric and hydrospheric evolution of the primitive earth. *Science* 160 (3829): 729–736.

Gilluly, J.; Waters, A. C.; and Woodford, A. C. 1968. Earthquakes and the earth's interior. In *Principles of geology,* 3rd ed. San Francisco: Freeman.

Mehlin, T. G. 1968. *Astronomy and the origin of the earth.* Dubuque, Iowa: Wm. C. Brown.

Press, Frank, and Siever, Raymond. 1974. *Earth.* San Francisco: Freeman.

Rubey, W. W. 1955. Development of the hydrosphere and atmosphere with special reference to the probable composition of the early atmosphere. *Geol. Soc. Amer.* Special Paper No. 62, pp. 631–650.

Scientific American. 1972. *Continents adrift* (Readings from *Scientific American*). San Francisco: Freeman.

4 The Beginning of Life

I am, in point of fact, a particularly haughty
and exclusive person of pre-Adamite ancestral descent.
You will understand this when I tell you that I can
trace my ancestry back to a protoplasmal,
primordial, atomic globule.

—W. S. Gilbert, *The Mikado*

THE RAW MATERIALS

The origin of life was perhaps the greatest of all events in biology. The processes involved in the event are many faceted and extraordinarily complex. When did life come into being, and precisely where? Chemically, what transformations and labyrinthic linkages were required? What, in the environment of a primordial earth festering with internal heat and bathed in lethal radiation, favored the miraculous transformation from nonliving to living?

Our planetary home is somewhat over 4.5 billion years old. The oldest fossil evidence of life thus far discovered occurs in rocks a little over three billion years old. These earliest organisms lived in a world largely devoid of atmospheric oxygen. They were anaerobic, and because they depended upon external organic molecules as food sources, they have been termed heterotrophic. Their environment lacked the shield of ozone which today is derived from oxygen and serves to absorb ultraviolet radiation from the sun. It is likely that life originated in such an environment. The event may have occurred in places protected from the deadly rays, or the organisms were by nature tolerant of this energy. Some believe that ultraviolet radiation may have impelled the molecules containing carbon to interact and produce organic compounds that eventuated in the evolution of life.

The elements needed to produce the organic compounds were available on the primitive earth. They consisted primarily of carbon, oxygen, hydrogen, and nitrogen. Falling rain probably gathered carbon dioxide from the air and washed it to the sea. Methane, ammonia, and phosphorus that remained from earlier atmospheres, or that were derived from contemporaneous volcanoes, were similarly dissolved and transported. But what was the manner in which these relatively simple materials developed into the complex components of a living thing?

Various experiments with mixtures of

compounds suggest plausible chemical pathways which might lead to life. To properly deal with the question, one must understand what constitutes life. Reduced to a fundamental definition, life is a system of four basic components. The first of these are proteins. Proteins are essentially strings of amino acids that are important in assembling many of the chemical compounds required for the production of nucleic acids. These nucleic acids comprise the second component of life. They are composed of strings of organic bases, five-carbon sugars, and phosphate units. Nucleic acids regulate the production of more protein and are believed to be important components of genes. Organic phosphorus compounds provide a third component of life. They function in transforming light or chemical fuel into the chemical energy required for the other reactions of the cell. The fourth essential for life is some sort of container such as a cell membrane. The enclosing membrane makes possible a relatively isolated chemical system and keeps the various compounds in close proximity to one another so they may interact.

Amino acids are the building blocks of proteins and vital to all living things. Two forces which existed in the early years of earth history may have been important in the development of amino acids. Prior to the complete formation of the ozone layer, ultraviolet rays streamed downward onto the earth's surface. This source of energy has been shown in the laboratory to be capable of tearing the atoms out of compounds such as water, ammonia, and hydrocarbons, and joining the atoms in the form of amino acids. A second energy source believed capable of performing a similar task might have been electrical discharges. Together lightning and ultraviolet radiation may have produced amino acids in the air, in tidal pools, in the surface layers of the sea, and wherever suitable quantities and proportions of raw materials existed.

SYNTHESIS OF VITAL MOLECULES

Is the general concept of the abiotic synthesis of amino acids really valid? The answer appears to be yes, for laboratory experiments within the last two decades have shown that mixtures of methane and ammonia can yield many different organic compounds under a variety of physical conditions. The main limitation to the success of the experiment appears to be the lack or near absence of free (uncombined) oxygen. Stanley Miller, at the suggestion of Harold Urey, performed the pioneer experiment. He infused an atmosphere like that of early earth's atmosphere into an apparatus similar to that shown in figure 4.1. As the mixture was circulated through the glass tubes, sparks of electricity were discharged into it as simulated lightning. At the end of only eight days, the condensed water in the apparatus had become turbid and

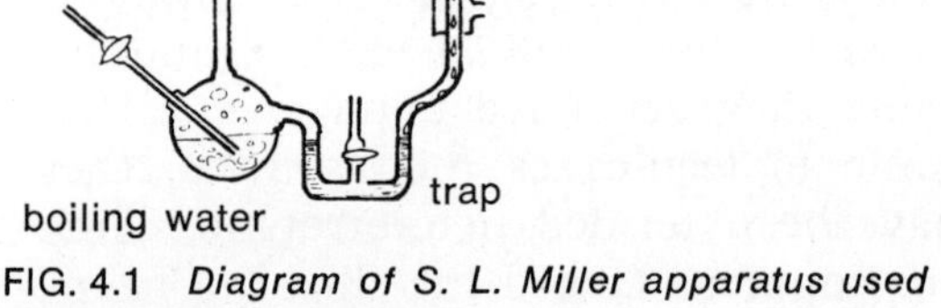

FIG. 4.1 *Diagram of S. L. Miller apparatus used to produce amino acids.*

deep red. Analysis of the crimson liquid showed that it contained significant quantities of a variety of amino acids, as well as more complex organic compounds that enter into the composition of all living things. In additional experiments by other biochemists, the ingredients, their proportions, and the stimuli were varied, yet they still confirmed the relative ease of artificially—abiotically—producing organic compounds. Some of the experimenters, doubting the need for a relative abundance of methane and ammonia, showed that organic compounds could be produced even from the gases of the second carbon dioxide, nitrogen, water vapor atmosphere. As a result of these experiments, it appears almost inevitable that amino acids would develop in the earth's prelife atmosphere or oceans. Because amino acids are very stable, one might also expect a gradually increasing abundance of these highly important building blocks of proteins.

For amino acids to form proteins, they must lose water. This can be accomplished experimentally by heating concentrations of amino acids to well over the boiling temperature of water (140° to 300° C.). However, the biochemist S. W. Fox discovered that the reaction occurred at temperatures as low as 70° C. if phosphoric acid was present. Fox was also able to produce proteinlike chains from a mixture of eighteen common amino acids. He called these chains *proteinoids* and reasoned that billions of years ago proteinoids may have been the transitional structures leading to true proteins. This is not extravagant conjecture, for Fox and his colleagues have found proteinoids similar to those he manufactured in his laboratory in the solidified lava and cinders adjacent to the vents of volcanoes. Apparently amino acids formed in the vapors of the volcano and were combined into proteinoids by the heat of the escaping gases.

Although long-chain nucleic acids have not as yet been experimentally produced under prelife conditions, some of the ingredients of nucleic acids have been formed spontaneously within mixtures of gases presumably present in the early atmosphere. It seems only a question of time before the method for artificially producing nucleic acids will be found.

Thus we see how it might have been possible for the pristine molecular chains and strings of life's components to have begun to form and accumulate. The chemical activity of this prelife would have gone unnoticed by man had he been present, but in the microcosmic world of tidal pools and wave washed shores, the molecules were nervously astir. One can visualize them forming aggregates small and large, the larger adding further molecules and growing in size and complexity, ascending gradually from nonliving, to transitional things of a biological netherworld, to the first living organisms. It could not occur in an age and an environment like our own, for oxygen and natural enemies would consume the delicate aggregates. But we are speaking of a time before there were organisms that cause decay, and before there was free oxygen.

SPECULATIONS ABOUT THE ORIGIN OF LIFE

Where did life originate? There are good reasons for thinking that it originated in the sea. The sea contains salts important to health and growth. The waters of the seas are universal solvents capable of dissolving a great variety of molecules, and able to ceaselessly circulate and mix them. Before life began, this constant motion would have favored frequent collisions of organic compounds, and thereby increased the probability of their joining together into larger and more complex structures. As the concentrations of mineral nutrients and carbon compounds increased, the seas became an ideal brew for the nurturing of pristine life.

Within the brew, the amino acids, as we have seen, would be all important. Under appropriate conditions they polymerize to form long chains which by definition are proteins. Proteins in turn, have a natural tendency to align themselves into still larger bodies of a size which identifies them as colloids. Finally, colloids have a number of properties that indicate their involvement in the origin of life. For example, they tend to attract water molecules and bind them securely. When colloids of opposite chemical charge collide, they sometimes combine their water envelopes to form spheres containing two or more colloidal particles. The resulting body appears enclosed in what might be similar to the precursor of the true cell membrane.

The Russian biochemist A. I. Oparin found that, in water solutions, certain large organic molecules attach themselves to specific other molecules so as to form larger aggregates that appear microscopically as tiny cell-like droplets. Oparin called these droplets *coacervates.* The electric charges of the molecules in each droplet cause the immediately adjacent water molecules to lie side by side like a fence. In this way, a membranelike wall of oriented water molecules is formed around the droplet. Such a wall may represent yet another path toward the development of the vitally necessary cell membrane. At any rate, the envelope of tightly bound molecules permits activities within the coacervates suggestive of the workings of a cell. It provides an enclosure in which chemical changes and new structural arrangements can occur. Molecules might move through the wall in either direction. Those moving inward constituted, in a primitive way, food for the growing coacervate. If the inwardly moving molecules exceeded those moving outward, the coacervate grew. If not, it was likely to disintegrate and, in a sense, "die." Long before the advent of the first amoeboid animal, "survival of the fittest" may have begun to operate among the coacervates. Those with more stable structures may have maintained themselves by consuming their less fortunate neighbors. It is not difficult to imagine, that on reaching some critical size and shape relative to the viscosity of its internal fluids, the coacervate might divide into two droplets, mimicking cell division. Those coacervates that were able to grow more rapidly and therefore split more often, received the lion's share of the available organic nutrients.

Another group of spherical bodies have been the object of particularly significant work by Dr. Sidney Fox. Fox discovered that when hot saturated solutions of proteinoids are cooled, the proteinoid molecules will aggregate and form large numbers of uniform, microscopic spheres. These cell-like bodies (microspheres) were remarkably stable and retained their general size and shape for several weeks. In general form, they were not unlike certain bacteria found in thermal springs. Furthermore, they exhibited osmotic shrinkage and swelling, budded in a manner somewhat similar to the budding in yeast, and even underwent septate division.

Thus it appears increasingly more plausible, that in the dynamic molecular brew of the primordial seas, the gap between living and nonliving was gradually closed. Pulsating molecular spheres cannibalistically contended with one another until eventually only those with the most stable and efficient metabolism remained. The victors, microscopic in size, but tested by the rigors of a completely unsympathetic environment, may have been the ancestors of us all.

THE NATURE OF EARLY LIFE

We can infer from our knowledge of modern biology, that the first living things were microscopic in size and unicelled. A perhaps too "advanced" living analogue might be the coccoid or spherical bacteria. It is likely that these earliest forms of life would not have evolved the means of manufacturing their own food, but fed upon the myriads of organic molecules in the sea. Today organisms living in this way are termed *heterotrophs.* The food gathered by these ancient heterotrophs was converted to energy necessary for vital functions. In the absence of free oxygen, there was only one way that the heterotrophs could have converted their food to energy, and this was by fermentation. There are many variations of the fermentation process. The familiar reaction of the fermentation of sugar by yeast, shown in simple form below, permits an organism to break organic molecules and rearrange their parts.

$$\underset{\text{sugar}}{C_6H_{12}O_6} \longrightarrow \underset{\text{carbon dioxide}}{2CO_2} + \underset{\text{alcohol}}{2C_2H_5OH} + \text{energy}$$

Animal cells also ferment sugar, yielding lactic acid rather than alcohol.

Unfortunately for the heterotrophs, but fortunate for the advance of life on the planet, the simple fermentation organism was to experience difficult times. By gobbling up the organic compounds of their environment, heterotrophs lived on borrowed time, eventually creating a food shortage which resulted in selective pressures for evolutionary change. At some point prior to the time that the food supply would have been seriously depleted, organisms evolved the ability to synthesize their needs from simple inorganic substances. These were the first autotrophic organisms. Unlike the heterotrophs, *autotrophs* are able to make their own food. The organisms that developed this ability saved themselves from starvation. Their evolution proceeded in diverse directions. Some autotrophs manufactured food from carbon dioxide and hydrogen sulphide. Presumably these were the ancestors of today's sulfur bacteria. Others (probably ancestors of nitrifying bacteria) utilized ammonia as a source of energy and matter. More important than these, however, was a group called photoautotrophs that could carry out photosynthesis. Their gift to life was the unique capability of dissociating carbon dioxide into carbon and free oxygen. The carbon was used to build more organic matter for the organism. The oxygen escaped to pave the way for yet another step in the evolution of primitive organisms.

A simplified photosynthesis reaction can be written as follows:

$$6CO_2 + 6H_2O \longrightarrow C_6H_{12}O_6 + 6O_2$$

Little by little the photoautotrophs expanded with the result that billions of tiny photoautotrophic oxygen generators became available on earth to change the primeval anoxygenic atmosphere to an oxygenic one. The change was awesome in its effect upon geological processes operating at the earth's surface and upon the subsequent evolution of life. Fortunately the change came slowly, for if oxygen had accumulated too fast, it would have been lethal to many of the early microorganisms. Some sort of oxygen acceptors were needed to act as safety valves. The crustal rocks, with their oxygen deficient iron minerals, provided suitable oxygen acceptors. Geologists have long been familiar with the great bodies of iron ore that characterize many Precambrian rocks. Often these rocks exhibit a banding that may represent a former fluctuating supply of oxygen from photoautotrophs that grew more or less luxuriantly during alternate periods of time. Above these banded rocks, more uniformly oxidized ores are often present, reflecting a more persistent presence of oxygen in the air. While the iron was blotting up excess oxygen, time was provided for the evolution of organisms with oxygen mediating enzymes. When most of the surficial iron had been oxidized, free oxygen began to accumulate in the atmosphere and hydrosphere. Solar radiation acted upon the oxygen to convert part of it to ozone, and the ozone in turn became an effective shield against ultraviolet radiation. The still primitive and vulnerable emerging organisms were protected, and the stage was set for the appearance of aerobic forms of life. Aerobic organisms use oxygen to burn their food. This method of providing energy was better than the earlier fermentation because it provided far more energy in relation to the food consumed. The reaction, which can be referred to as cold combustion, can be simply written as follows:

$$6O_2 + C_6H_{12}O_6 \longrightarrow 6CO_2 + 6H_2O + \text{energy}$$

It is likely that the greater amount of energy now available to organisms was an important factor in the evolution of larger and far more complex forms of life.

It has become rather commonplace in classrooms to hear teachers use the expression "from amoeba to man" when considering the totality of evolutionary history. The phrase is an unfortunate one, for it contains the implication that the already highly organized and complicated protozoans represent a first step in the history of life. Many biologists believe that the developmental changes leading up to that first amoeba were far more intricate and extraordinary than all of succeeding evolutionary history. The first phenomenal step on the journey to the amoeba was the fabrication of something living from lifeless organic molecules. These organisms we have called heterotrophs. Their successors, the photosynthesizing autotrophs, have actually been found preserved as fossils. The earliest of these organisms, like the modern bacteria and blue-green algae, are relatively simple forms which have neither a cell nucleus nor special internal organelles. They are termed procaryotes (fig. 4.2) and are incapable of either body cell division (mitosis) or germ cell division (meiosis). Nevertheless, they represent an important second step forward in the history of primordial life. The procaryotes, possibly because they are essentially sexless and thus restricted in the amount of variability they can develop, have shown great stability and conservation through some two billion years of earth's history. Evolution proceeded from these procaryotes to organisms with a definite nuclear wall, well-defined chromosomes, and the capacity for meiosis and sexual reproduction. These were the eucaryotes (fig. 4.2). With their immense capability

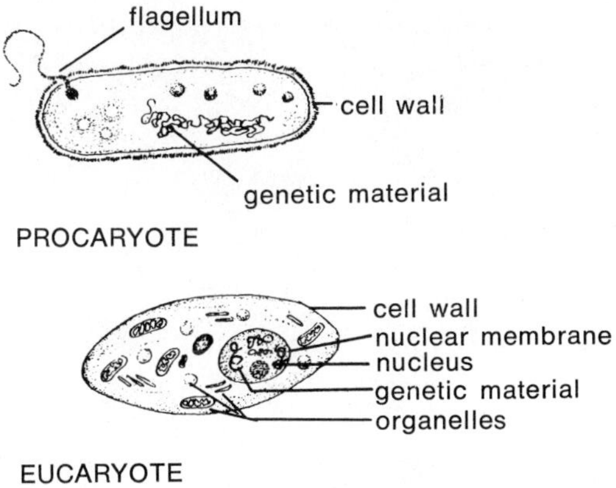

FIG. 4.2 *Comparison of procaryotic and eucaryotic cells.*

for genetic diversity, the eucaryotes provided the third great step forward. Thus those who write the history of life "from amoeba to man" should beware, for their volume shall lack important early chapters.

SUMMARY

The earliest forms of life on earth were probably anaerobic (able to exist without oxygen). They were either capable of surviving under ultraviolet radiation, or they developed in places shielded from such radiation. Their origin involved the combining of carbon, oxygen, hydrogen, nitrogen, and atoms of certain other elements into simple organic molecules such as amino acids. Amino acids are essential components of proteins. Proteins in turn are required for the production of nucleic acids, another ingredient of living things. The origin of life, or simply biogenesis, also required the acquisition of organic phosphorous compounds and the evolution of an enclosing membrane for the first "cell." Experiments like the one performed by Stanley Miller have shown that amino acids can be synthesized under laboratory conditions that approximately duplicate those on earth over three billion years ago.

The energy required to convert atoms to simple molecules, and simple molecules to complex giant molecules could have come from the heat extruded from the earth itself, from the sun's radiation, or from the electrical discharges of lightning. Gradually, over millions of years, the organic compounds accumulated in the earth's water bodies. Chemical combinations of the smaller molecules produced increasingly larger and more complicated ones. Presumably, the most advanced of these eventually acquired some of the most elementary characteristics of a living cell. These first hypothetical organisms were consumers, living off the chemical energy of the smaller molecules they devoured. With the evolution of photosynthesis, life was able to use light energy from the sun to manufacture food. Free oxygen accumulated as a by-product of photosynthesis. This made possible the evolution of organisms that use oxygen to burn their food for energy. It made respiration possible.

SUPPLEMENTARY READINGS AND REFERENCES

Adler, Irving. 1957. *How life began.* New York: John Day.

Cloud, P., and Gibor, A. 1970. The oxygen cycle. *Scientific American* 273 (3): 110–123.

Fox, S. W. 1960. How did life begin? *Science* 132 (3421): 200–208.

Fox, S. W., and Dose, K. 1972. *Molecular evolution and the origin of life.* San Francisco: Freeman.

Keosian, J. 1968. *The origin of life.* 2nd ed. New York: Reinhold Book Corp.

Miller, S. L. 1960. Formation of organic compounds on the primitive earth. In *Aspects of the origin of life* pp. 85–97. Oxford: Pergamon Press.

Oparin, A. I. 1959. *The origin of life on earth.* Oxford: Pergamon Press.

Ross, H. H. 1966. *Understanding evolution.* Englewood Cliffs, N.J.: Prentice-Hall.

Rutten, M. G. 1971. *The origin of life by natural causes.* Amsterdam: Elsevier Books.

Wald, G. 1964. The origins of life. *Proc. Nat'l. Acad. Sci.* 62:595–611.

5 Vestiges in Chert

Somewhere, somehow, sometime, in the mysterious chemistry of carbon, the long march toward the talking animal had begun.
—Loren Eiseley, *The Immense Journey*

THE OLDEST FOSSILS

It was only two decades ago that fossils from rocks more than a billion years old were unknown. This was not for lack of searching. Whenever the opportunity arose, geologists patiently scrutinized the ancient rocks in the hope of obtaining a measure of the fame that would come to the discoverer of the earth's oldest traces of life. Unfortunately, they were largely unsuccessful, and geology students were told that early forms of life probably had unpreservable soft bodies. Anyway, they would have been destroyed during rock erosion or deformation. However, geology teachers have learned to be both cautious and optimistic, and lectures usually closed with the prediction that fossils billions of years old did exist somewhere and someday would be discovered.

At least in part, the lack of success resulted from the fact that searchers did not concentrate on the right kind of rocks. More often than not, the sediment in which the ancient fossils were entombed was later subjected to heat and pressure or chemical alteration. Vast numbers of Precambrian fossils would have been obliterated. However, among Precambrian rocks, chert, because of its hardness and chemical stability, often escaped substantial alteration. As a result, primordial clues to life can be found sealed in a cherty mass precipitated from Precambrian waters over three billion years ago.

Chert is a common rock composed of the elements silicon and oxygen. Its principal mineral component is a variety of quartz known as chalcedony. Flint is the popular name for a black variety of chert in which the dark color is due to inclusions of organic matter. The origin of chert is a complex problem. It also appears that different varieties of chert are formed by somewhat different processes. Some cherts are replacements of earlier carbonate rocks or sediments. Others appear to have formed from the solution and reprecipitation of silica from organisms having siliceous skeletal parts. Some cherts apparently formed from sil-

From *The Immense Journey* by Loren Eiseley with the author's permission. Published by Random House Inc., 1957.

ica precipitated directly from aqueous solution. Many of the earliest bacterial and algal remains of living things float in a matrix of chert, and appear to have been trapped in a siliceous gel that later turned to chert in much the same way as modern animals can be artificially preserved in plastic. Recently, laboratory simulations of the geochemical processes involved were successfully performed by William Schopf and John Oehler at the University of California, Los Angeles. Modern blue-green algae were stirred into a solution of colloidal silica, the colloid was converted to a solid gel by treatment with salt and hydrochloric acid, and then transformed to synthetic quartz by elevating the temperature and pressure for a few weeks. The "artificial microfossils" produced closely resembled their fossil counterparts in morphology, amount of alteration, and color.

Among the oldest cherts in the world are those located in the border area between the Republic of South Africa and Swaziland. There, near the town of Barberton, a now famous cherty formation is exposed. Its name, somehow reminiscent of another story of genesis, is the Fig Tree Formation. The Fig Tree rocks are composed of variously colored cherts, ironstones, slates, and indurated sandstones. By means of radioactive isotopes, the formation has been dated at 3.1 billion years old, but sedimentation and the actual entrapment of the now fossilized organisms probably occurred about 3.2 billion years ago. Specimens of the chert were originally studied by Harvard University paleobotanist Elso S. Barghoorn and his former graduate student J. William Schopf. Using the techniques of electron microscopy, Barghoorn and Schopf discovered a number of tiny rod-shaped structures about five microns in length and two microns in diameter (fig. 5.1). A double-layered cell wall much like that seen in many modern bacteria surrounded the bodies. So as to draw attention to their great antiquity and single

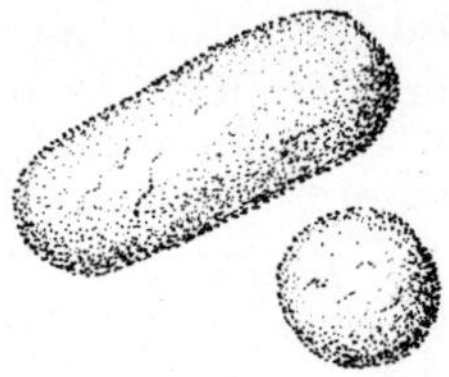

FIG. 5.1 Eobacterium, *the "dawn" bacterium, about 0.6 microns long. (Adapted from E. S. Barghoorn, 1971, "The Oldest Fossils,"* Scientific American, *p. 31.)*

cell habit, Barghoorn and Schopf named these tiny fossils *Eobacterium isolatum.* In a search for somewhat larger remains, thin sections of the chert were prepared and examined with the optical microscope. To the delight of the researchers, they were able to find numerous spheroidal fossils very similar to certain of the blue-green algae. Indeed, they may even represent the ancestral stock of these primitive plants. They have been given the name *Archaeosphaeroides barbertonensis* (fig. 5.2). Along with *Eobacterium isolatum,* they bear testimony to the fact that the transition from nonliving to liv-

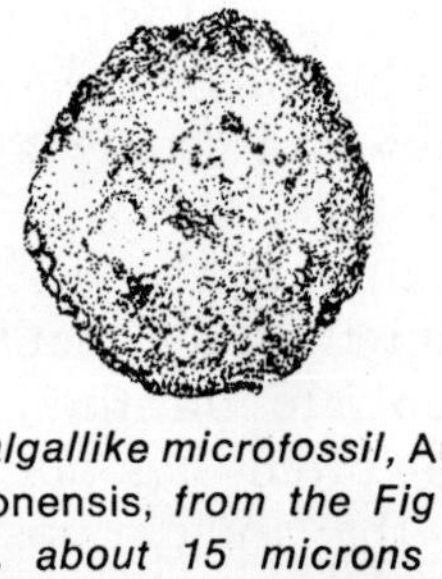

FIG. 5.2 *The algallike microfossil,* Archaeosphaeroides barbertonensis, *from the Fig Tree Formation of Africa, about 15 microns in diameter. (Adapted from a photograph, courtesy of J. Wm. Schopf.)*

ing had been achieved at some point in time prior to 3.2 billion years ago.

DIVERSIFICATION BEGINS

Evolution teaches us that when a few relatively unspecialized organisms find themselves in an environment which contains unoccupied niches and few predatory enemies, they may experience a relatively rapid increase in numbers and diversity. This exploitation of many new evolutionary opportunities has been termed "explosive evolution" by some students of evolution. The term is, of course, misleading, for untold numbers of decades are involved in these changes. Regardless of the suitability of the term, however, evidence for such an episode of diversification can be found in the black cherts of the Gunflint Formation on the northern shore of Lake Superior. Radiometric age determinations indicate this formation is approximately 1.9 billion years old. It contains a varied and abundant flora of the so-called thread bacteria and nostocolean blue-green algae which may be generally referred to the lower thallophytes. Unbranched filaments, some of which are septate, have been given the name *Gunflintia* (fig. 5.3) and are probably related to similar appearing modern blue-green algae. Among the

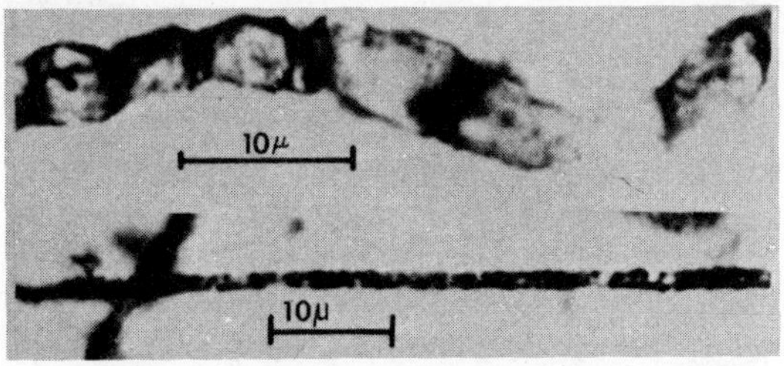

FIG. 5.3 *The Precambrian filamentous algae,* Gunflintia grandis *(above) and* Gunflintia minuta *(below). (Courtesy of J. Wm. Schopf.)*

more finely septate forms, those named *Animikiea* (fig. 5.4) exhibit many parallels with such living blue-green algae as *Oscillatoria* and *Lyngbya*. Other fossils

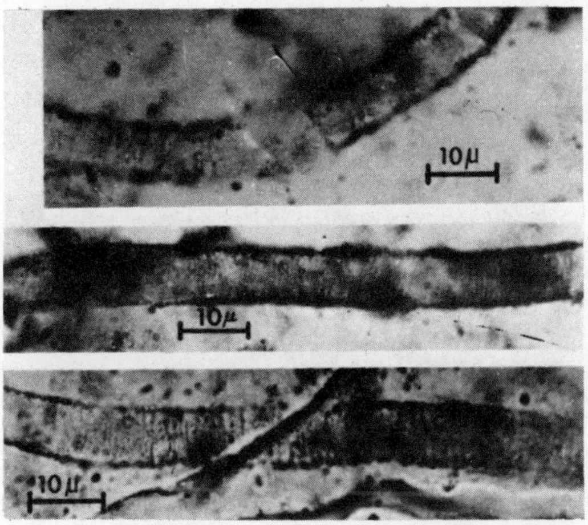

FIG. 5.4 Animikiea septata, *a filamentous microorganism from the Gunflint Chert. (Courtesy of J. Wm. Schopf.)*

such as *Eoastrion* (the "dawn star") resemble living iron and magnesium reducing bacteria. *Kakabekia* and *Eosphaera* (figs. 5.5 and 5.6) are sufficiently different from any known organism that their classification is uncertain. That some of these and other organisms of the Gunflint were actively producing oxygen and changing the composition of the atmosphere does seem certain. Not only do many of them resemble living photosynthetic organisms, but the enclosing chert contains pristane and phytane—hydrocarbons regarded as breakdown products of chlorophyll. In addition, small laminated domes and pillars in the Gunflint resemble structures produced today by shallow water, dome-building, photosynthetic algae.

The record continues to improve as one next examines the cherts of the Bitter Springs Formation of Central Australia. These one-billion-year-old cherts contain colonial microfossils that grew in laminar sheets and mats near the surface

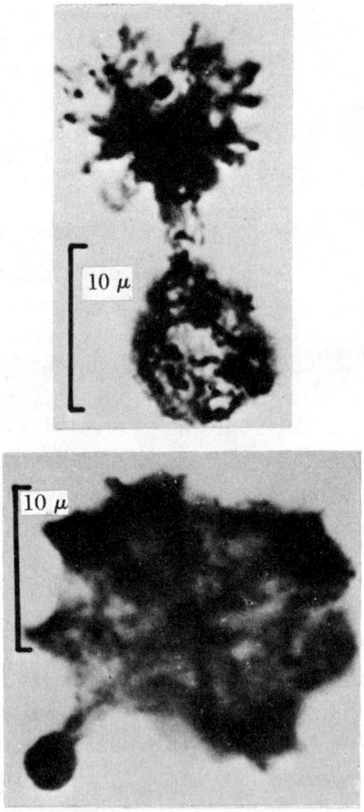

FIG. 5.5 *Two stages in the life history of* Kakabekia umbellata *(Courtesy of J. Wm. Schopf.)*

of shallow marine waters. Thirty distinct microfossils have been identified as part of the Bitter Springs Flora. About half of these are blue-green algae similar to species living today. Also included are fossils of colonial bacteria, funguslike filaments, and spheroidal green algae. The green algae are particularly significant, for preserved within them are distinct cell nuclei, indicating that the passage from procaryotic to eucaryotic organisms had been accomplished at least as long ago as one billion years.

The laminated structures characteristic of the Bitter Springs and Gunflint

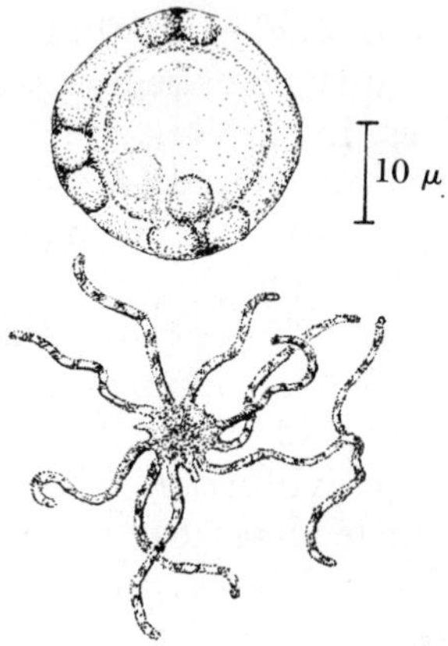

FIG. 5.6 Eosphaera tyleri *(above) and* Eoastrion bifurcatum *(below). (After E. S. Barghoorn, 1971, "The Oldest Fossils,"* Scientific American, *p. 36.)*

Formations are called stromatolites (fig. 5.7). They occur frequently in Late Precambrian sedimentary rocks and are the most abundant of Precambrian fossils. Stromatolites are most often found in limestones, dolostones, iron formations, or in former carbonate rocks that have been replaced with silica. The laminar structure is always well developed and there are wide variations in size and

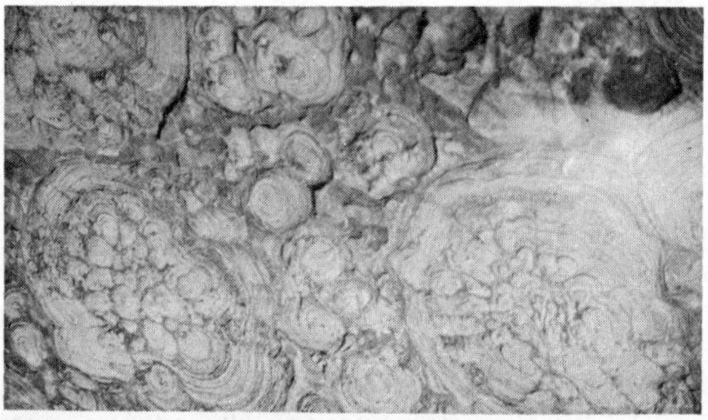

FIG. 5.7 *Stromatolites. (Courtesy of G. W. Stose, U.S. Geologic Survey.)*

shape. They may be massive, frondose, digitigrade, or incrusting. Present-day colonies of blue-green algae form similar laminated structures. These algae cause the development of stromatolites by trapping fine particles of sediment between the minute filaments of their matlike colonies. The carbonate is temporarily bound within a film of gelatinous organic matter. Successive additional layers result in the laminations. It is interesting to note that stromatolites have been found in rocks of Rhodesia that are more than two billion years old.

DAWN ANIMALS

Sooner or later the assiduous search for the remains of more advanced, multicellular forms of Precambrian life was certain to produce positive results. Indeed, since 1961 a fascinating assortment of Precambrian fossils have been found in Australia, Siberia, and southwest Africa. The fossils are primarily impressions in sedimentary rocks of the types of animals that can be loosely grouped under the name "metazoans." Metazoans are multicellular, possess more than one kind of cell, and have the different cells organized into tissues or organs. The best known fossil assemblage of early soft-bodied metazoans was chiseled from the Pound Quartzite—a formation exposed in the Ediacara Hills, about two hundred eighty miles north of Adelaide, South Australia. Of the over six hundred specimens collected, all represented soft-bodied animals. They included several different kinds of jellyfish, soft corals resembling the modern sea pen *Pennatula*, polychaete worms, echinoderms, arthropodlike animals, and a number of unique forms whose precise biological affinity is uncertain. The Ediacara fauna occurs in rocks whose estimated age is about seven hundred million years old. Its importance lies in the fact that it provides us with a glimpse of the ancestors of the vast array of invertebrates that seem to appear quite suddenly on the scene one hundred million years later at the beginning of the Paleozoic Era.

The Fig Tree, Gunflint, and Pound formations provide isolated glimpses of the progress of evolution in the long succession of years prior to the Paleozoic Era. It is as if we were travelers on a fictional time train, and only at chance moments were we permitted to raise the window shades and view the antique landscape. Life, of course, is a continuous progression, and the rock strata of intervening ages can be expected to yield additional fossils. Indeed, such strata are currently being closely scrutinized for transitional organisms—the missing links in the chain of life. Figure 5.8 illustrates the current state of our knowledge and lists the more important fossiliferous formations. One notes in examining this chart that, until very late in the Precambrian, fossile metazoans are lacking. Although many explanations have been advanced to account for this paucity of older Precambrian metazoans, it seems likely that these creatures could not begin their expansion until the atmosphere contained sufficient oxygen, perhaps as much as 3 percent of the present oxygen content. This amount would have been sufficient for metazoans of low oxygen requirements to survive. Possibly at this point, metazoan evolutionary radiation began, and subsequently the record improved. Henceforth, our story will involve more fact and less speculation.

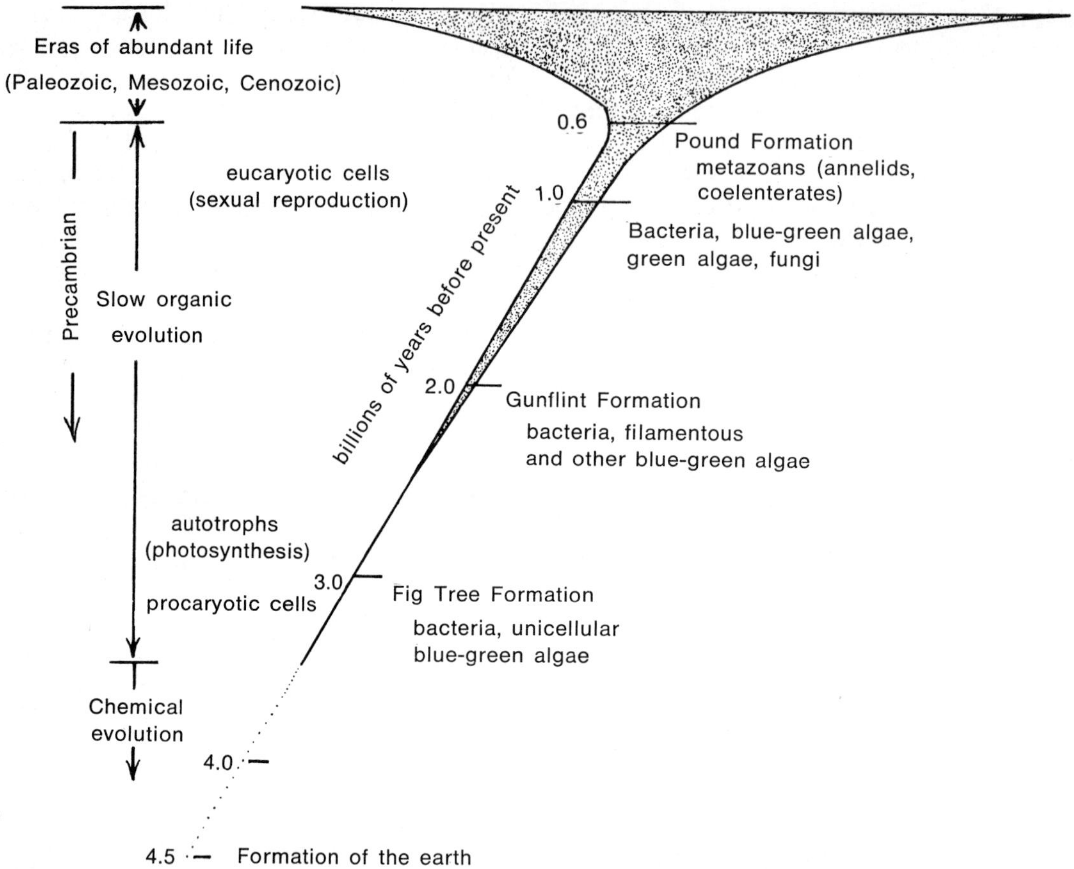

FIG. 5.8 *Timetable for early earth history. (Modified from E. S. Barghoorn, 1971, "The Oldest Fossils,"* Scientific American, *p. 41.)*

SUMMARY

The world's oldest fossils have been found in sedimentary rocks of the Fig Tree Formation. The best evidence for the age of this formation indicates that it is somewhat in excess of 3.2 billion years old. Chert, a submicrocrystalline form of quartz, is a characteristic lithology of the Fig Tree Formation. Indeed, nearly all the Precambrian microscopic organisms discovered in recent years have been found in chert deposits. Refinements in sample preparation and electron microscopy have made possible the examination of fossil bacteria less than a micron in size. Such bacteria, usually in the form of tiny rods, occur along with filamentous organic structures in the Fig Tree. Also present are hydrocarbons known only to be formed by life processes.

A more diverse group of ancient microorganisms has been found in samples of the now famous Gunflint Chert which is exposed along the shores of Lake Superior in western Ontario. The Gunflint is about a billion years younger than the Fig Tree sediments. The most abundant

Gunflint fossils are microscopic unbranched filaments, some of which are septate, and most of which are related to modern blue-green algae. Certain of the filaments contain spheroidal sporelike bodies resembling spores contained in living algae and iron bacteria. Spheroidal and star-shaped forms also occur not unlike certain bacteria found today. Interestingly, the formation also contains hydrocarbons believed to be the breakdown products of chlorophyll.

Thirty new species of microfossils including the remains of colonial bacteria, blue-green algae, and green algae are now known from the billion-year-old Bitter Springs Formation of Australia. The microfossils in the Bitter Springs are found in laminated structures called stromatolites and include funguslike organisms, blue-green algae, and spheroidal green algae. Stromatolites are frequently found in Precambrian and more recent sedimentary sections. The oldest stromatolites occur in the Bulawayan Limestones of Rhodesia and indicate oxygen production through photosynthesis approximately 2.5 billion years ago.

Although tracks and burrows presumed to have been made by metazoans (many-celled animals) are known from rocks somewhat over a billion years old, fossils showing body form have not been found in rocks older than about 0.7 billion years, or very latest Precambrian. The best-known discovery consists of coelenterates, annelids, and arthropods impressed into the Pound Quartzite of Ediacara Hills, South Australia.

SUPPLEMENTARY READINGS AND REFERENCES

Barghoorn, E. S. 1971. The oldest fossils. *Scientific American* 244 (5): 30–42.

Barghoorn, E. S., and Tyler, S. A. 1965. Microorganisms of the Gunflint Chert. *Science* 147 (3658): 563–577.

Barghoorn, E. S., and Schopf, J. W. 1966. Microorganisms three billion years old from the Precambrian of South Africa. *Science* 152 (3723): 758–763.

Glaessner, M. F. 1961. Precambrian animals. *Scientific American* 204 (3): 72–78.

———. 1966. Precambrian paleontology. *Earth Science Reviews* 1: 29–50.

Schopf, J. W. 1970. Precambrian microorganisms and evolutionary events prior to the origin of vascular plants. *Biological Reviews* 45 (3): 319–352.

Schopf, J. W., and Blacic, J. M. 1971. New microorganisms from the Bitter Springs formation (Late Precambrian) of the North-central Amadeus Basin, Australia. *Jour. Paleo.* 45 (6): 925–960.

6 The World of Plants

We propose in these pages to direct our attention to the vegetable world, on the existence of which that of animals is dependent.

—Alexander von Humboldt, *Aspects of Nature*

INVASION OF THE LAND

Before there were animals on earth, there were plants. In the preceding pages, we noted that the fossil record of plants begins with Precambrian siliceous rocks of Africa and North America. That primeval and microscopic flora consisted largely of blue-green algae, colonial bacteria, funguslike organisms, and filamentous or spheroidal green algae. A few other unicellular forms of uncertain affinities were also found. The bacteria and blue-green algae are, of course, what we might have anticipated finding in the earliest fossil record of plants. Also significant are the green algae or Chlorophyta, for this phylum contains organisms that are biochemically more advanced than the blue-green algae. Because of their advanced attributes, the Chlorophyta are generally considered ancestral to the Phylum Tracheophyta. This great phylum is composed of the enormous population of vascular land plants, including ferns, trees, and shrubs that clothe our continents. The relationship of tracheophytes to chlorophytes is suggested by the observation that both phyla contain the same sort of green pigment, and they produce the same variety of carbohydrate during photosynthesis. Also, some members of the Chlorophyta have actually become adapted to fresh water—a necessary accomplishment in the vegetative invasion of the continents. Today, some green algae even live quite satisfactorily in moist areas on land.

The evolution of vascular land plants involved the development of familiar plant structures for obtaining and transmitting water and nutrients. Surficial moisture in terrestrial environments is an undependable source of water. However, there is a somewhat greater persistence of moisture beneath the surface in the pore spaces of soil. Many nonvascular photosynthetic plants, however, are unable to use subsurface water because to do so they would also have to live in the underground, and there would be no light for photosynthesis. The trend in evolution, therefore, was toward plants with roots for obtaining water and nutrients, a vascular system for conducting fluids through the body of the plant, and chlorophyll-bearing leaves to receive these nutrients and accomplish photosynthesis. No less important was the eventual evolution of special structures for the protection of the gametes.

Just as there are missing links in the animal world, so also do such gaps occur in the known historical record of plants. Fossils of plants transitional between green algae and the first unquestioned tracheophytes are as yet undiscovered. We may surmise that the intermediate stage was represented by a plant somewhat better adapted to life on land than most green algae, but still lacking the vascular tissue for conducting fluids, and also devoid of specialized structures for protecting the gametes against loss of vital moisture. Today liverworts and mosses have some of the attributes one might expect to see in the hypothetical transitional plant. However neither of these groups is considered to be in the direct line of evolution toward higher land plants. As nearly as can be ascertained from the sequence of fossils, mosses developed too late to be ancestral to tracheophytes. They have survived down to the present day by remaining relatively small and clinging to moist environments.

The problems to be overcome in plant colonization of the land were formidable, and the event was late in coming. The first unquestioned remains of vascular plants are found in rocks of Upper Silurian age. These are the Psilopsida. The psilopsids were primitive plants, most less than a foot tall. Structurally, they consisted of horizontal stalks or rhizomes which grew in the damp soil just below the surface of the ground. Along the rhizomes were short vertical "stems" bearing smaller branches and spore sacs. Paleobotanists are intrigued by the presence of delicate strands of wood cells which extend up through the center of the branches. Although psilopsids had neither true roots nor leaves, they were, nevertheless, more advanced than the algae or mosses in that they did have some woody tissue. No less important was the rudimentary vascular system which permitted part of the plant to exist underground where there was water but little light, and another part to grow where water supplies might be uncertain, but sunlight assured. The psilopsids paved the way for the evolution of large upright trees. Because of the development of wood, plants were able to stand tall against the pull of gravity and resist the force of high winds. By late in the Devonian, there were forests of lofty, well-rooted, leafy trees. One of these forests has left ample fossilized remains in the vicinity of Gilboa, New York. The fossilized trunks of some of the Gilboa trees are more than a meter in diameter. Of course, these were not psilopsids, but more advanced plants called lycopsids that we will discuss next.

FOR BETTER PROPAGATION

When one views the entire span of plant history, three major episodes are clearly apparent. Each of these involved the development of plants with increasingly more effective reproductive adaptations. The first episode led to seedless plants that were more advanced than the psilopsids. These had true leaves and extensive root systems for anchoring and absorption. These more progressive, seedless, spore-bearing plants include the well-known ferns, and less familiar lycopsids and sphenopsids that grew in the extensive coal swamps of the Late Paleozoic. The second episode of plant evolution saw the advent of a large group of nonflowering, pollen and seed-producing plants. The appearance of plants with both seeds and flowers—the angiosperms—completed the three stages of plant progression.

More Spore Producers

Among the distinctive seedless plants of the coal swamp flora were the scale trees or lycopsids. Today they are represented by a group of small survivors, the club mosses, of which the ground pine *Lycopodium* is a member. Small size, however, was not particularly characteristic of Late Paleozoic lycopsids. Some grew to lofty heights. The trunk of *Lepidodendron* regularly exceeded thirty meters in length and was crowned with a skyward reaching series of forked branches. The slender leaves of the scale trees emerged directly from the trunks and, after being dropped, left a regular pattern of leaf scars. In *Sigillaria,* these scars are arranged in vertical rows, whereas the diamond-shaped scars of *Lepidodendron* spiral diagonally (fig. 6.1).

Jointed seedless plants called sphenopsids grew side by side with the lycopsids in the coal swamps. Living sphenopsids include the scouring rushes and so-called "horsetail fern." Sphenopsids, as exemplified by fossil *Calamites* and *Annularia* (fig. 6.1), have longitudinally ribbed stems that are divided at regular intervals by transverse joints. Clusters of leaves or smaller branches emerge at these joints. In sphenopsids, the spores develop in special sacs at the tips of the stems.

True ferns, members of the Phylum Pteropsida, grew alongside the scale trees. Many of these ferns were tall enough to qualify as trees. Like the lycopsids and sphenopsids, they reproduced by means of spores, and these were carried in regular patterns on the undersides of the leaves.

Seeds and Pollen

In the next great event of the history of plants, forms with seeds and pollen, but not flowers, began to spread across the continents. Such plants are known informally as "gymnosperms." Their new system of propagation involving seeds and pollen enhanced the possibilities for efficient reproduction in dryer terrains. The gymnosperms began their evolution in Late Paleozoic and became the dominant plants of the Mesozoic. They were the ancestral stock from which higher seed-bearing plants emerged. The four major groups of gymnosperms are the seed ferns or Pteridospermophyta (now totally extinct), the cycadophytes, the ginkgoes, and the conifers.

Seed ferns had fernlike leaves, but unlike true ferns, they reproduced by seeds borne on the leaves. *Neuropteris* and *Pecopteris* (fig. 6.1), when found in association with the seed structures, are generally considered valid members of the Pteridospermophyta. Another ancient seed fern of the Southern Hemisphere is *Glossopteris.* The tongue-shaped leaves of *Glossopteris* are familiar to geologists arguing the merits of the theory of continental drift. The geographic distribution of the *Glossopteris* flora provides one line of evidence in favor of the former existence of a now fragmented supercontinent. According to theory, the parts of that supercontinent have drifted on huge

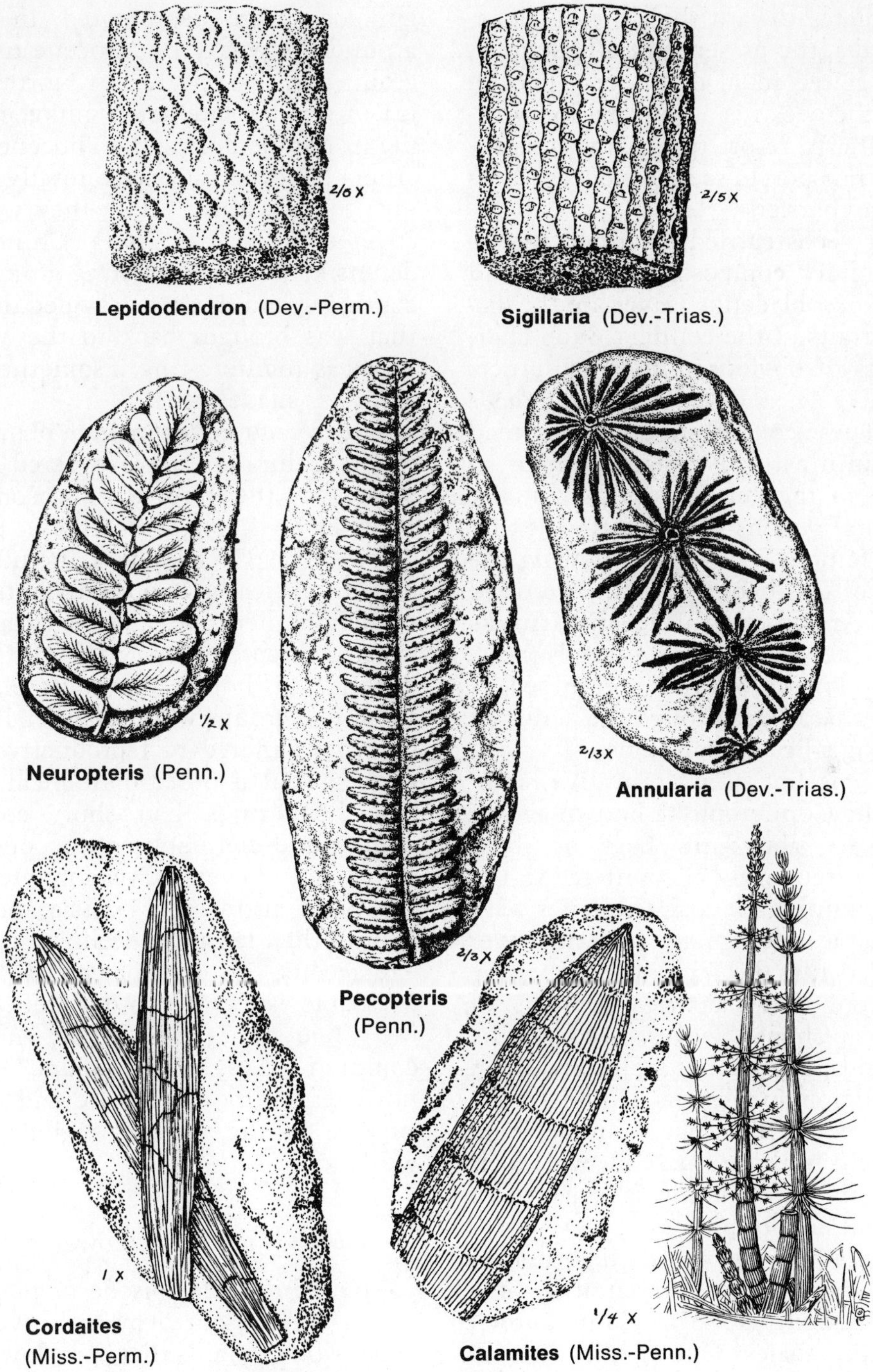

FIG. 6.1 *Plants of the coal swamps.*

crustal plates toward their present locations. Thus, the presence of *Glossopteris* and associated fossil plants on opposite sides of the ocean is accounted for without having to resort to explanations involving transport of seeds or spores across vast ocean barriers.

Naked seeds carried in cones, trunks rather solidly composed of wood, and needlelike or bladelike leaves are the distinctive traits of the conifers. Such characteristics are obvious in the pine, spruce, hemlock, yew, and redwood of today's forests. These cone-bearing plants spread widely during the Permian, perhaps in response to the general uplift of many continental regions that occurred at that time. A famous locality for viewing the remains of fossil conifers is the Petrified Forest of Arizona. At this locality, trunks and branches were buried in sands transported by Triassic streams, and these logs were later petrified through the action of infiltrating siliceous solutions.

In the coal measures, primitive members of the Coniferophyta known as cordaites were abundant. Logs of these plants are frequently encountered in the course of mining for coal. Cordaites were mostly large trees, many towering over fifty meters into the sky. Their branching limbs were crowned with growths of large, strap-shaped leaves. Seed-bearing shoots and pollen cones were carried on the smaller branches. As would be expected in coniferlike plants, ample wood was present in the trunk. However, their more primitive condition is reflected in the central core of pith.

Only a single species, *Ginkgo biloba,* survives today as a representative of a once flourishing group of plants known as the Ginkgoales. During the interval between Permian and Early Cenozoic, ginkgoes were represented in forests around the world. By Miocene time, the group had disappeared in North America. A similar fate befell ginkgoes growing in Europe during the Pliocene. Westerners believed they were totally extinct until, in the early 1700s, they were discovered in eastern China. Chinese gardeners had been cultivating ginkgoes for centuries, and thereby provided the stock that was brought back to the Western world as an interesting, if sometimes malodorous, shade tree.

The cycadophytes include plants with rough, columnar stems covered by leaf bases of the previous seasons and crowned by tough, leathery, pinnate leaves. The Cycadophyta includes two paleontologically important groups, the extinct Bennettitales and the cycads. The latter are represented today by the sago "palms." Although the adult plants in these two groups were superficially similar, they differed in reproductive structures as well as other anatomical details. The bennettites had short columnar trunks and hermaphroditic cones that resembled flowers. Cycadophytes were abundant during the Triassic and Jurassic and then began to decline during the Cretaceous. To the paleobotanist, the Mesozoic is more appropriately named the "Age of Cycads," than the more common "Age of Dinosaurs." Today, however, only a few cycad genera survive in the warmer regions of the world.

Plants with Flowers

The third great episode of plant evolution is represented by the advent and expansion of plants that bore flowers and carried enclosed seeds. Such flowering plants, or angiosperms, have further improved the reproductive method. The trend toward greater efficiency in reproduction began, as we noted earlier, with the spore-bearing, seedless psilopsids, ferns, and lycopsids. In those groups, the adult plant, or sporophyte, produces special reproductive cells that do not require fertilization. These reproductive cells are the spores. When the spores are dropped by the sporophyte onto the moist ground, they develop into small leafless plants called gametophytes. Gametophytes in turn produce egg and sperm. The mobile sperm must swim a small distance through a film of moisture to an egg, which, on fertilization, develops into the larger, leafy parent sporophyte. Obviously, there is an element of risk involved in the hazardous journey of the sperm. Dehydration of spore, egg, or sperm is an abiding danger. The mechanisms of evolution worked toward more efficient methods. An improvement came with the development of seeds. The early seed plants did not have their seeds enclosed by a covering or by soft tissue as, for example, the seeds in an apple or melon. Indeed, their name "gymnosperm" means "naked seed." The gymnospermal plants retained the special gametophyte stage of their seedless predecessors, but not as an independent plantlet. Instead, small male and female gametophytes develop within the moist tissues of the parent (sporophyte) plant. The female gametophyte develops into seed-producing organs, and the male into bodies capable of producing sperm-carrying pollen grains. These tiny pollen grains are highly resistant to drying, and are readily transported by wind or animals. So resistant are they, that their undecomposed coverings can be extracted from rock strata so as to provide information about climates of long ago.

In the reproductive sequence of the seed-bearing but flowerless plants, the pollen is first transported to the female reproductive organ. On arrival, they grow a slender tube that reaches to the egg-bearing structure of the female gametophyte. The sperm then travels in relative safety down the moist tube to fertilize the egg. The dangerous overland swim is eliminated. In the next stage of the sequence, the fertilized egg develops into a seed. The seed, essentially a dormant embryonic sporophyte, is surrounded by starchy nutrients. When the seed is transported to an environment favorable for its germination, it sprouts to become a new sporophyte plant.

In angiosperms (the term means "covered seeds"), the seeds develop within the capsular ovule chamber of the flower. Organs within the flower (the stamens) produce pollen. The pollen is transported to yet another organ of the flower called the pistil. As in gymnosperms, a pollen tube is produced from the pollen grain after it has come to rest on a pistil. After the pollen tube has penetrated downward into the ovule chamber, the sperm moves through the tube to the egg where fertilization takes place. The seed then begins its development. The colored petals of flowers are really modified leaves and are useful in attracting insect and bird pollenators. The enclosed seed of the angiosperm permits the growth of edible coverings. These nutritious fruits are eaten by animals, and the resistant seeds dispersed after being voided. One has only to examine one's trouser legs after a walk in the woods in the late summer or early fall to learn of the many other hook and burrlike devices, and floating tufts that other angiosperms have evolved for seed dispersal.

Angiosperm pollen grains extracted from Jurassic rocks provide the earliest evidence for the existence of flowering plants. By Cretaceous time, angiosperms were conspicuously present. Forested areas included stands of magnolia, poplar, sassafras, birch, and willow (fig. 6.2). By the end of the Mesozoic Era, angiosperms clearly surpassed the nonflowering plants. Their diversity and range of adaptation has been truly incredible. Angiosperm trees, shrubs, vines, herbs, and grasses gave a modern aspect to the landscape and, either directly or indirectly, affected the evolution of insects, birds, and mammals.

SEA PLANTS

Life in the oceans, as on land, is dependent on plants. The record of marine plants extends far back into the Precambrian Era. As discussed earlier, those ancient plants included bacteria, blue-green algae, and green algae. Red and green algae capable of causing the precipitation of calcium carbonate have been noted in rocks as old as Ordovician. Brown algae, of which the living kelp is an example, have left uncertain imprints in Devonian rocks and may have existed much earlier.

Two phyla of marine plants are well known as fossils, and also of exceptional importance in food production in the sea. These are the Pyrrophyta, or dinoflagellates, and the Chrysophyta. The latter phylum includes diatoms and coccolithophorids. Species of these two phyla are components of the great phytoplankton pastures of the seas. On these and associated plants depend the oceans' zooplankton, from which in turn extend food chains to fish, whales, seabirds, and ultimately man, dining elegantly on lobster at Delmonico's. The phytoplankton began a rapid expansion during the Cretaceous. Earliest fossil coccolithophorids and diatoms are found in rocks of Jurassic age. Dinoflagellate evolution may have begun in Silurian time with a group of possible pyrrophytes called hystrichospherids.

Preservation of the outer coverings of coccolithophorids and diatoms is improved because of their mineral composi-

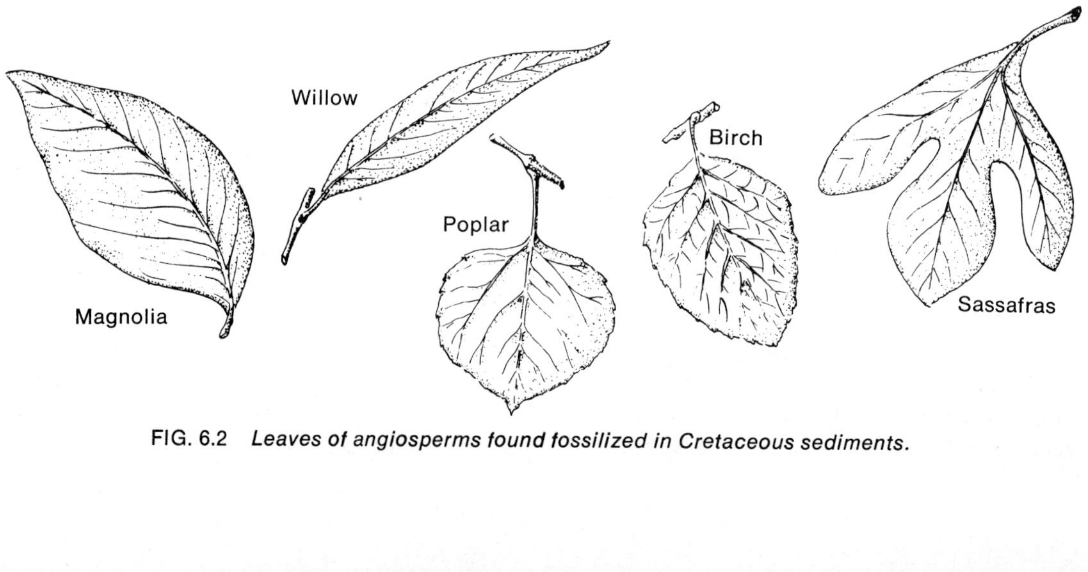

FIG. 6.2 *Leaves of angiosperms found fossilized in Cretaceous sediments.*

tion. Coccolithophorids build spherical coverings (coccospheres) of calcium carbonate plates (fig. 6.3). The plates are usually circular, elliptical, or pentagonal and are themselves composed of still smaller crystallites arranged in a uniform manner. Diatoms (fig. 6.4) secrete siliceous coverings which are formed like medicine capsules with two slightly overlapping halves. The calcareous and siliceous oozes that blanket extensive areas of the ocean floors are composed largely of the remains of these two groups of phytoplankton. (The protozoan radiolaria and foraminifera also contribute to the formation of these organic oozes.) Coccoliths are a major constituent of the rock called chalk. All of these groups of microfossils are used in dating and correlating marine strata penetrated in the course of drilling for oil.

FIG. 6.3 *A coccolith, one of many platelike structures covering a coccosphere.*

Plants, whether they comprise myriads of microscopic phytoplankton of the oceans or vast fields of grain on our continents, sustain us all. Our reliance on plants is not confined to those presently living, for plants of bygone eras have provided us with coal and oil to fuel our factories and warm our homes. The very air we breathe contains its vital gift of

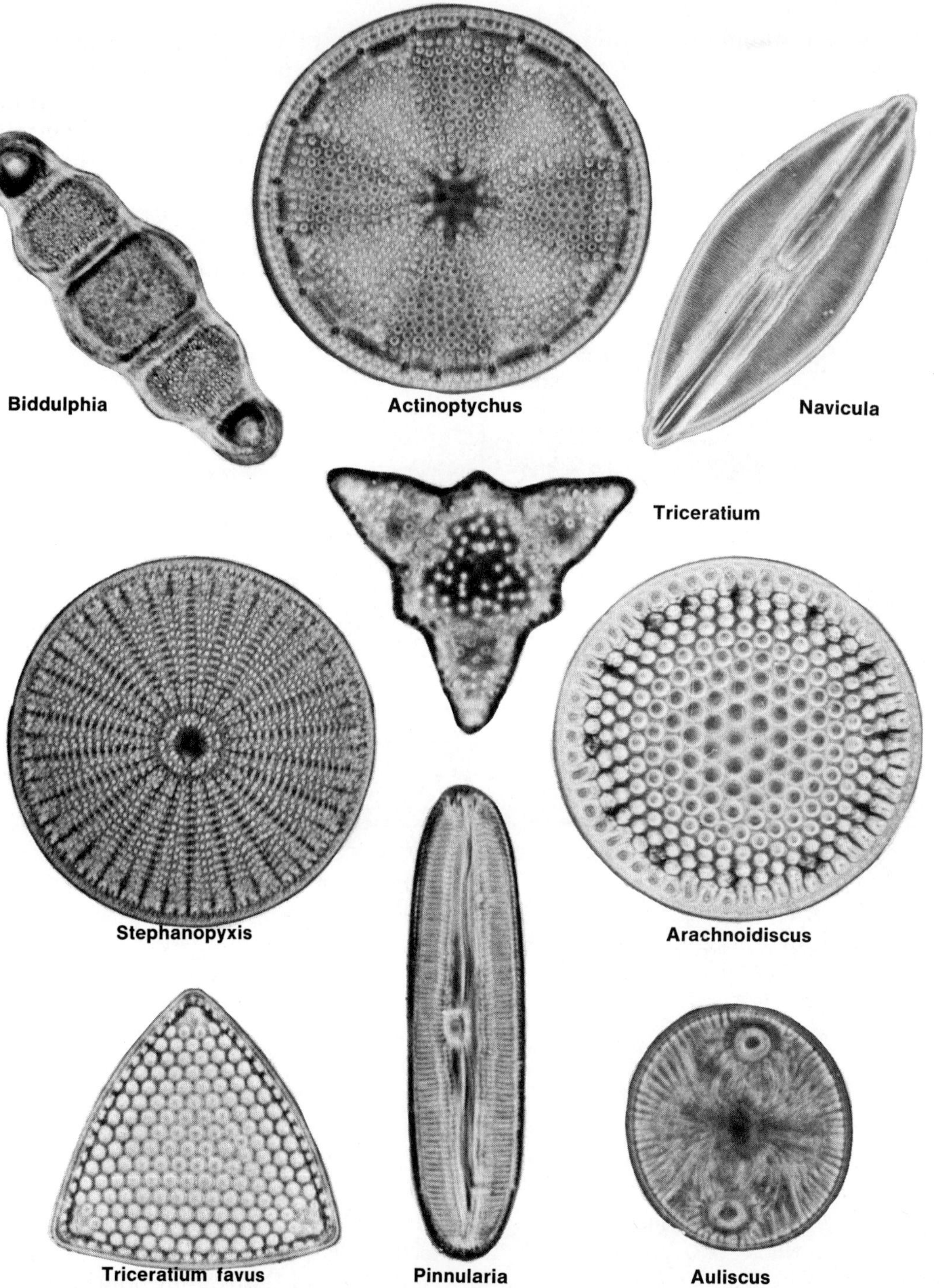

FIG. 6.4 *Diatoms.*

oxygen, produced in the leaves of green plants, and within algal cells. Not at all frivolous were the once popular bumper sticker signs which asked, "Have you thanked a green plant today?"

SUMMARY

The fossil record for plants began over three billion years ago with the discovery of simple filamentous forms that bear a resemblance to certain modern algae and fungi. In general, Precambrian plant life appears to have consisted entirely of nonvascular forms that lived in aquatic habitats. The concentrically laminated deposits known as stromatolites first appear in the Precambrian, although they are found in younger rocks as well. Stromatolites are quite similar to layered structures formed today by blue-green algae. Fossil evidence of Chlorophyta or green algae also occurs in the Precambrian. These finds are noteworthy, for the Chlorophyta may have been the predecessors of vascular land plants.

The advent of the vascular land plants involved the evolution of familiar plant structures for obtaining and transmitting water and nutrients. The first of these vascular plants were the leafless psilopsids of the Late Silurian. These were soon to be joined by lycopsid scale trees, those relatives of the horsetails called sphenopsids, and the true ferns. All of these seedless groups were abundantly represented in the great coal forests of the Late Paleozoic.

Plants that produce seeds have been reported as fossils from rocks of Mississippian age, although they probably had evolved from fernlike plants during the preceding Devonian. The seed ferns or pteridosperms were very common during the Late Paleozoic. Also, before the end of the era, conifers had become well established. Although the lycopsids and sphenopsids declined during the Mesozoic Era, the seed plants continued to spread. The foliage of Cycadales and superficially similar Cycadeoidales (Bennettitales) is so frequently fossilized in Mesozoic beds that the era has been termed the "Age of Cycads." Ginkgoes were contemporaries of the cycads in forests around the world.

The angiosperms, or flowering plants, made their appearance during the Mesozoic and expanded dramatically during the last period of that era. As they rose to dominance, gymnospermous plants like the cycads gradually diminished in numbers and variety. Forests containing most of the genera of hardwood trees familiar to us today were already present near the end of the Cretaceous. As a result, the landscapes of the following Cenozoic Era had a distinctly modern appearance. Perhaps the single most important botanical event of the Cenozoic was the spread of the grasses. Actually, the oldest grass seeds found as fossils occur in Late Cretaceous rocks. They expanded steadily, until by Miocene time extensive prairies had developed. The feeding habits of many of the worlds herbivorous mammals evolved in response to the spread of this important food source.

In the seas and oceans, mostly microscopic plants flourished during the Mesozoic and Cenozoic Eras. Diatoms and coccolithophorids made their appearance in the Jurassic and have been especially abundant in Cenozoic marine strata. Today, the empty skeletons of these tiny plants carpet the seafloor with oozes. They have been a basic food source in the marine world for over a hundred million years.

SUPPLEMENTARY READINGS AND REFERENCES

Andrews, C. A. 1961. *Studies in paleobotany.* New York: Wiley.

Delevoryas, T. 1962. *Morphology and evolution of fossil plants.* New York: Holt, Rinehart and Winston.

———. 1966. *Plant diversification.* New York: Holt, Rinehart and Winston.

7 A Multitude of Shellfish

The great heart of nature beats, its throbbing stimulates the pulse of life, and not until that heart is stilled forever will the rhythmic tide of evolution cease to flow.

—Richard Swann Lull

THE RECORD

The earth's first animals were invertebrates. By comparing them with their still living relatives, we know that they lived in the oceans. The tree of life itself appears to be rooted in seawater. The ocean provided nutrients, moderated temperatures, shielded delicate tissue from lethal rays, and generally protected life in its feeble beginnings.

If one surveys the broad panorama of marine invertebrate life over the past six hundred million years, certain general observations become apparent. One notes, for example, that the principal phyla appeared in the first period of the Paleozoic Era (fig. 7.1). Indeed, as the pa-

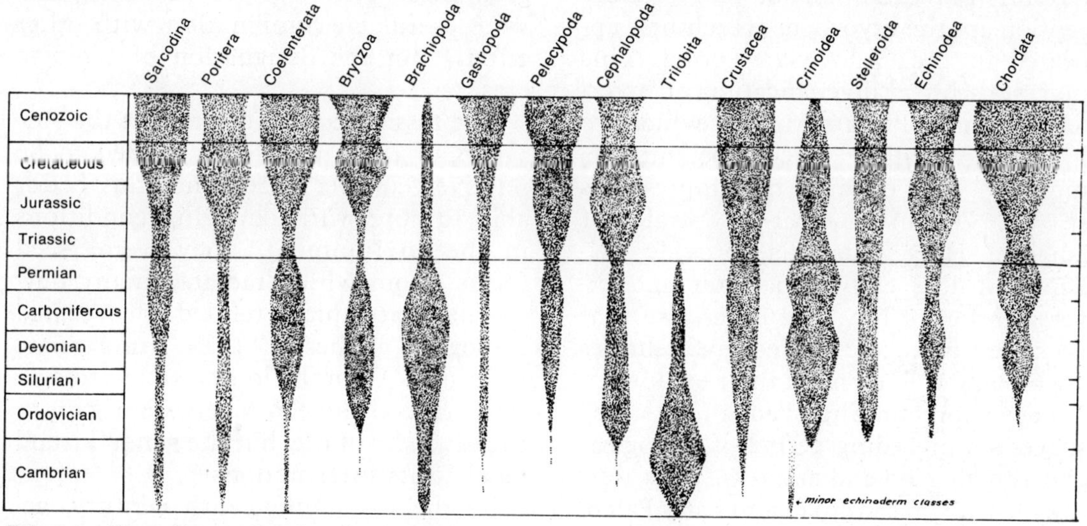

FIG. 7.1 *Geologic ranges of major categories of animals. Thickness indicates relative abundance. (Modified from A. L. McAlester, 1968,* History of Life, *Prentice-Hall, Inc., pp. 54 and 55.)*

leontologist T. de Chardin has suggested, "Life no sooner began, than it swarmed." However, to some extent this impression of sudden evolution may be a misconception resulting from the capriciousness of the fossil record. The abrupt appearance of fossiliferous strata does not necessarily require an equally hasty appearance of animals, but perhaps only the spread of the shell-building habit with its excellent potential for producing fossils. Shells do have some advantages. They provide support and protection for soft tissues. In particular environmental situations, the development of shells might be favored by selection. If the advantage provided by the trait was profound, it could be expected to spread rapidly and thereby account for the sudden appearance of fossil remains.

Another observation obtained from the fossil record is that all of the phyla of invertebrate animals with preservable hard parts are of great antiquity. Most appear to have originated at or near the Cambrian-Precambrian boundary. An exception are the bryozoans which first appeared in the Ordovician Period. This suggests a broad diversification of invertebrates in the Precambrian for which we have little fossil evidence. Coelenterate, mollusc, and echinoderm impressions in the 700-million-year-old strata of Ediacara Hills, Australia, provide evidence of this early expansion and diversification of life. One notes also that the oldest Paleozoic shelled invertebrates are already well along in their evolutionary development. Their "advanced state" suggests a preceding period of biological experimentation and progress. Thus, during the Late Precambrian and Early Paleozoic, the great phyla had arisen. Later diversification of life occurred within the major phyla, rather than varying to such a degree as to require the taxonomic formulation of a new phyla.

Yet a third observation can be made from the historical record of life. Although there has been a persistent increase in numbers of different species, these trends have been upset or even reversed at times for one or more groups as a result of environmental adversity. Such crises were usually accompanied by widespread extinctions. Long ago, while engaged in the study of fossils from vertical successions of strata, geologists noticed these relatively short episodes of widespread extinctions which interrupted the long periods of more or less orderly evolution. Indeed, these interruptions helped to provide the temporal boundaries for the geologic time scale. Minor extinctions, such as those that occurred at the family taxonomic level were taken to indicate the termination of geologic periods. Mass extinctions involved a large number of different and larger taxonomic groupings. These greater catastrophies were useful, in combination with other criteria, for the designation of geologic eras.

The fossil record suggests that the survivors of these calamities tended to be relatively unspecialized and thus better able to cope with changing conditions in the environment. They were "seed forms" from which radiated many new species that rapidly refilled the vacated ecological niches of their unfortunate precursors. Over the long span of geologic time, there has been a persistent gain in the diversity of life, but the general trend has had its interruptions.

By the early nineteenth century, geologists in France and England had recognized another useful characteristic of the fossil record. They noted that the overall characteristics of the marine invertebrate faunas of each of the geologic eras tended to be distinctly different from the faunas and floras of older or younger eras. Such dissimilarities provided these early investigators with an additional stratigraphic tool for recognizing and correlating particular sequences of strata. Trilobites, brachiopods, nautiloid cephalopods, tabulate and rugose corals, and certain bryozoans were abundant in Paleozoic rocks. Remains of hexacorals, pelecypods, echinoids, and ammonoids characterized the beds of the Mesozoic. Although the Cenozoic faunas had a certain overall similarity to the Mesozoic, its distinctness was apparent in the total lack of ammonoids, and the presence of contemporary foraminiferal families, naked cephalopods, and a host of modern molluscans and echinoids.

CATEGORIES OF THE FREQUENTLY FOSSILIZED

We have briefly noted some of the overall characteristics of marine invertebrate history, and may now turn to a discussion of the eight phyla having abundant fossil representatives. These are the Sarcodina, Porifera, Coelenterata, Bryozoa, Brachiopoda, Mollusca, Arthropoda, and Echinodermata.

Sarcodina

Like the well-known amoeba (fig. 7.2), the sarcodinids are creatures which do not have differentiation of their bodies into separate cells. Their life processes are carried out within a single protoplasmic vessel analogous to a single cell of multicellular, or metazoan, animals. Because

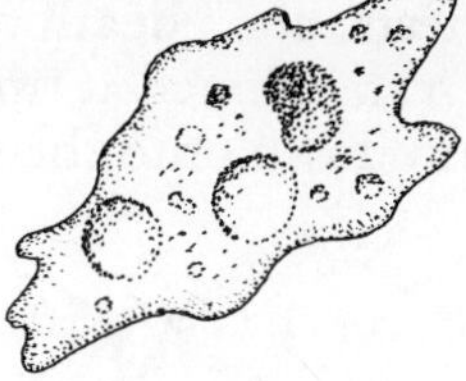

FIG. 7.2 *Amoeba.*

such organization can function only in minute animals, sarcodinids are generally microscopic in size. They are further characterized by a changeable cell form, and the ability to construct mobile extensions of the body which are called pseudopodia. Although sarcodinids are simpler animals than metazoans, they are still extraordinarily complex.

Modern classifications place the Phylum Sarcodina within the Kingdom Protista, which includes organisms that are neither distinctly animal nor plant. In older classifications, sarcodinids were placed in one of the four member classes of the Phylum Protozoa. Among the sarcodinids are two orders that are capable of secreting intricate and often beautiful supportive structures composed of calcium carbonate or silica. One of these orders, and the most important in paleontology, is the Order Foraminiferida (fig. 7.3). Because of their small size, their remains can be obtained intact even from the broken pieces of rock recovered in the course of drilling for oil. "Forams" are nearly all marine organisms that build their tests by adding chambers in one or more rows, in coils, or in spirals. They take their name from the large number of perforations (foramina) in the test or skeleton. A larger opening, the aperture, is also present. Most species of foraminifers are bottom dwellers (benthonic). Fewer species, but large populations, float near the surface of the open oceans as plankton. Their tests, upon death of the organism or when abandoned at times of reproduction, rain down onto the seafloor like

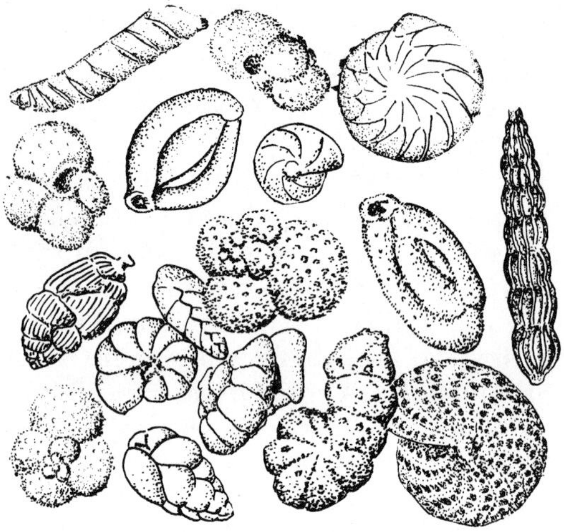

FIG. 7.3 *Foraminifers.*

continuous snowfall. Among the planktonic foraminifers, a common form is *Globigerina* (fig. 7.4). The extensive, limey, deep sea sediments known as globigerina ooze contain in great abundance the tests of *Globigerina* and related forms. The remains of *Globigerina,* along with calcium carbonate remains of other planktonic organisms, form a common variety of limestone known as chalk.

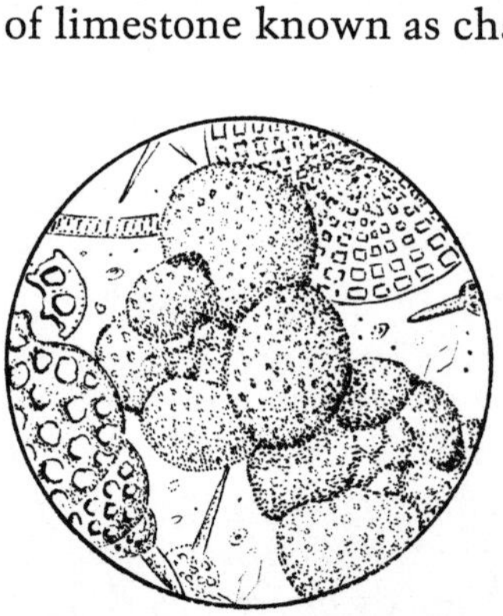

FIG. 7.4 *Test of a species of* Globigerina *(center).*

The oldest foraminifers are simple forms found in rocks of the Early Paleozoic. The order did not become particularly varied or abundant until Late Paleozoic when families evolved which were about the size and shape of wheat grains. These were the numerous, diverse, and internally complex foraminifers loosely termed fusulinids. In some limestones, fusulinids comprise the entire bulk of the rock. Even when less abundant, they are useful index fossils for correlation of strata.

Foraminifers having a more modern appearance became abundant during the Mesozoic. It was in the Cretaceous Period that planktonic forms appeared and expanded rapidly. Like their Paleozoic ancestors, their remains provide excellent index fossils and are extensively used in petroleum exploration. Because they are sensitive indicators of water temperature and salinity, they have provided data useful in the study of ancient marine environments all over the world. Certain species obtained from cores of deep sea sediment are providing interesting measures of areal and temporal fluctuations of glacial climates during the Pleistocene Ice Age. In these studies, the temperature requirements of the foraminifers found in successive layers of the cores can be determined by direct comparison with living species. One can also study the isotopic composition of the calcium carbonate secreted by the once living animal. The ratio of isotopes varies according to the temperature of the water at the time the animal lived. It has been discovered that the ratio between the isotopes oxygen-18 and oxygen-16 in the tests decreases 0.02 percent for every 1°C. decrease in water temperature. Oceanic temperatures that prevailed many hundreds of thousands of years ago can be ascertained using these methods.

Radiolarians are sarcodinids with threadlike, radially directed pseudopodia which project from beautifully filigreed, latticelike opaline or proteinaceous skeletons (fig. 7.5). In some regions of the sea,

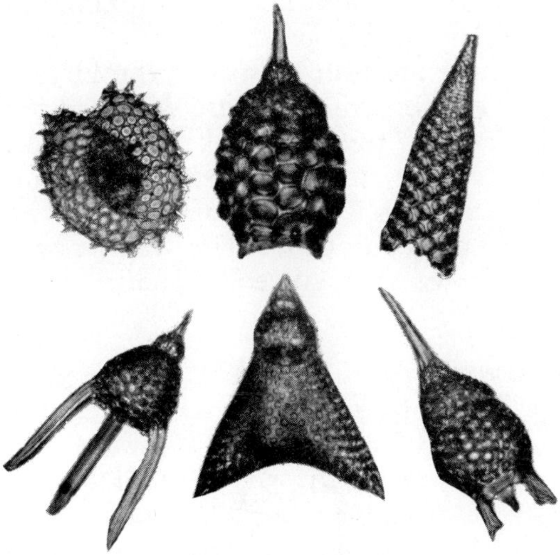

FIG. 7.5 *Radiolaria.*

radiolarian tests accumulate in deposits called radiolarian ooze. These oozes occur frequently in deeper or colder parts of the oceans. Because of the increased concentration of carbon dioxide in the colder waters, calcium carbonate tests of foraminifera often dissolve as they settle to the seafloor, or the initial secretion of a calcite test is inhibited. However, the siliceous radiolarians are unaffected and tend to accumulate. Radiolarians have been reported in rocks as old as Precambrian, but occur with greater certainty from the Devonian to the present.

Porifera

Porifera are sponges. Sponges appear to have evolved from colonial flagellate protistids. This was an important transition and bears directly on the origin of the metazoans. Evidence indicating the precise path by which the changes were accomplished is vague and may well be lost in a hiatus resulting from unfossilized ancestors. Although there are several conflicting hypotheses, one called the "colonial theory" is usually favored because of its simplicity. The colonial theory proposes that an aggregate of acellular protistid cells eventually achieved a division of functions, so that the colony began to operate as a unit. Beginning with the initial aggregate, those cells located around the outside would develop nervous and locomotor functions, whereas the more deeply embedded cells might become specialized in food gathering, digestion, and reproduction. While colonial flagellate protistids do not exhibit this specialization of function, simple metazoans like sponges show it well. Also, sponges have a rather characteristic type of cell called a choanocyte which has a nearly identical counterpart in the flagellates.

Among the invertebrates, sponges comprise a long-lived and relatively conservative evolutionary side branch. They have been in existence since the Precambrian. Fossil sponges like *Astraeospongia* of the Silurian, the Devonian glass sponge *Hydnoceras,* and the peculiar Pennsylvanian *Girtyocoelia* are common index fossils (fig. 7.6). All but one group of sponges are marine, and they are found in all the oceans from shallow water to the deepest abyss. The walls and cell layers of most sponges are supported against deformation by skeletal elements. These may be spicules of varying form and mineral composition, or fibers of proteinaceous matter termed spongin. The spicules are important in the preservation and later identification of fossil sponges. In general, sponges with calcium carbonate spicules require shallower, warmer waters. The siliceous spicule producing "glass sponges" are tolerant of deeper water and colder temperatures.

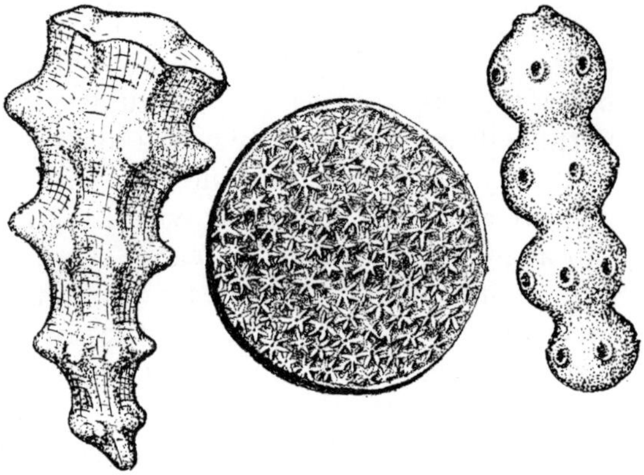

FIG. 7.6 *Fossil sponges, from left to right,* Hydnoceras, Astraeospongia, *and* Girtyocoelia.

Although sponges vary widely in size and form, their basic plan (fig. 7.7) is that of a highly perforated vase with walls modified by folds and canals. The body is attached to the seafloor at the base, and an excurrent opening, or osculum, occurs near the top. There are structurally three layers in the wall. Facing the internal space is a layer of choanocytes, or collar cells. Externally there is a layer of flat cells which function as a sort of protective pavement. Between these layers we find a nonliving gelatinous substance (mesenchyme) through which discrete spicule producing amoeboid cells move about.

It appears that the truly diagnostic characteristics of the Phylum Porifera are negative. That is to say, they lack the usual metazoan attributes. Sponges are without organs and lack even a true gut. Water currents moving through the sponge are created by pressure differences within the interior cavities which are the result of the beat of choanocyte flagellae. Suspended food particles carried into the sponge by water currents are ingested by the choanocytes and then undergo intracellular digestion such as is found in unicellular colonial protistids.

There are four classes of the Phylum Porifera. In general, these can be differentiated on the basis of skeletal composition and spicular architecture. Skeletons of the Class Desmospongea consist entirely or in part of spongin which may be reinforced with siliceous spicules whose

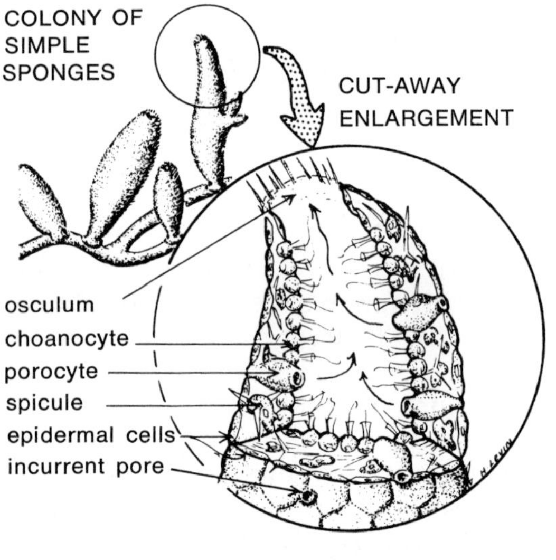

FIG. 7.7 *Features of a simple sponge.*

rays meet at 60 or 120 degree angles. The Hyalospongea (as implied by their name, *hyalos* = glass) develop siliceous spicules whose rays generally meet at right angles. Sponges characterized by spicules made of calcium carbonate constitute the Calcispongea (*calx* = lime). All three of the above classes have been in existence since Cambrian time, and the Hyalospongea may have been present even earlier. A new class of sponges, the Sclerospongiae, has recently been proposed for a group living in reefs off Jamaica and composed of organic fibers, siliceous spicules, and the calcium carbonate mineral aragonite.

Archaeocyatha

The Archaeocyatha ("ancient cups") are a puzzling group of Cambrian organisms that constructed calcareous, cone, or vase-shaped skeletons about eight or ten centimeters long. Because the cones had perforated double walls separated by septalike partitions, the archaeocyathids are sometimes presumed to have affinities to both the sponges and the corals. In other features, however, these fossils are sufficiently different to merit being placed in a phylum of their own. Archaeocyathids were the first major phylum to become extinct. They are important as the earliest of the reef-building animals. These creatures appear to have lived along the coast lines of warm shallow Cambrian seas. The Australian archaeocyathid limestones are approximately sixty meters thick and extend in a narrow band for over two hundred kilometers. Similar, but smaller, archaeocyathid "gardens" occur in North America and Siberia.

Coelenterata

The Phylum Coelenterata is known for the beauty and great diversity of its members. Living coelenterates (fig. 7.8) are represented by sea anemones, corals, sea fans, jellyfish, and the tiny hydra of school biology classes. All coelenterates

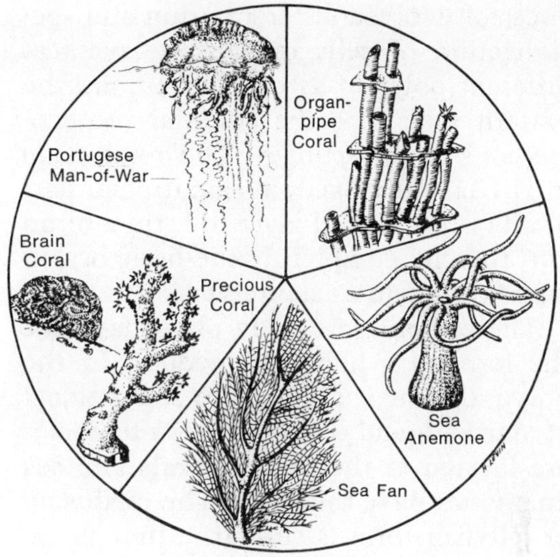

FIG. 7.8 *Diversity of living coelenterates.*

are aquatic, and most are marine. They are characterized by a saclike body with a mouth surrounded by tentacles at one end. Cnidoblast cells containing special stinging capsules called nematocysts assist the coelenterate in capturing food. When a small worm or crustacean touches one of the coelenterate's tentacles, the unlucky victim is suddenly riddled with a burst of paralyzing, barbed threads.

There is considerable difference of opinion as to the evolutionary path leading to the coelenterates. The paleontologist James R. Beerbower (1968) provides an interesting synthesis of earlier works on colenterate origins. He traces the hypothetical route of evolution through four stages. Initially, benthonic, colonial, vagrant protozoans develop. These then give rise to creatures with a ventral feeding and locomotive layer, a protective dorsal layer, some anterior-posterior differentiation for directional movement, and an interior cell mass for vital functions. In the third stage, the hypothetical ancestor evolved differentiation and specialization of cells in various layers at different poles so as to approximate the pattern seen in some colonial protozoans and larval sponges. The final stage might have involved invagination of part or all of the ventral layer to form a blind gut, thereby completing the basic organization seen in coelenterates.

The coelenterate body plan may take the form of a polyp or medusa. In the polyp, the sac is attached to some object at one end, and the mouth and tentacles are located at the other. Corals and sea anemones have this form. The medusoid or jellyfish form is similar, although inverted so that the mouth and tentacles lie below the sac (fig. 7.9).

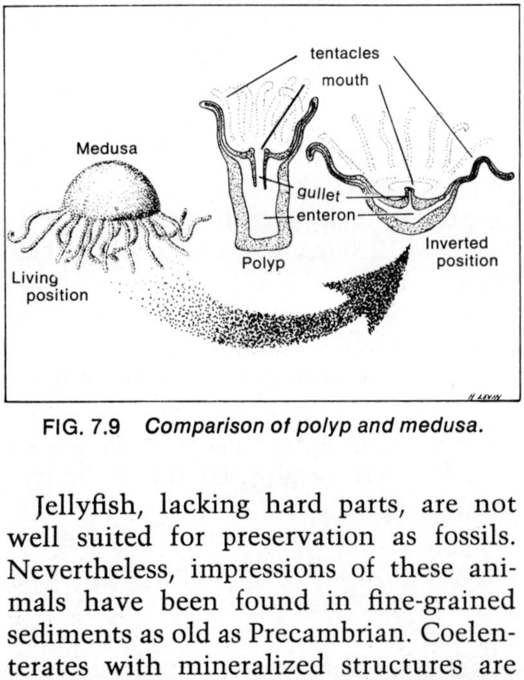

FIG. 7.9 *Comparison of polyp and medusa.*

Jellyfish, lacking hard parts, are not well suited for preservation as fossils. Nevertheless, impressions of these animals have been found in fine-grained sediments as old as Precambrian. Coelenterates with mineralized structures are abundant in the fossil record. Today such animals build extensive reefs in warm, shallow parts of the oceans. By implication, fossil coral reefs are believed to have formed in similar environments and are thus useful in making paleoclimatological interpretations. Some of the more primitive of the stony coelenterates were the stromatoporoids (fig. 7.10). During

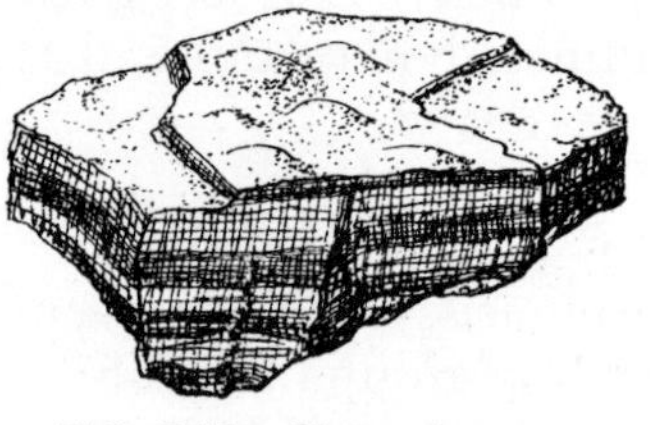

FIG. 7.10 Stromatopora.

the Silurian and Devonian, stromatoporoids built their thinly laminated structures into extensive reefs. The stony corals of the Class Anthozoa have always been the most important of fossil coelenterates. In these animals, the polyp secretes a calcareous cup in which it resides, and from which it grows. The coral may live alone or combine with others to form colonial masses of enormous size. The cup, or theca as it is called, may be divided by vertical plates named septa which mark the position of deep folds in the soft body wall. Septa lend support to the polyp and separate the layers of tissue which provide greater digestive area for the organism. As the animal grows, it secretes horizontal plates called tabulae (fig. 7.11). Corals are identified and classified according to the characteristics of the tabulae, the septa, and lesser skeletal features. For example, the Paleozoic rugose, or tetracorals, insert septa at four locations during adult growth and thus have septa in multiples of four. In other Paleozoic corals, septa are absent or poorly developed so that tabulae become the primary feature in classification. These are the tabulate corals, and they

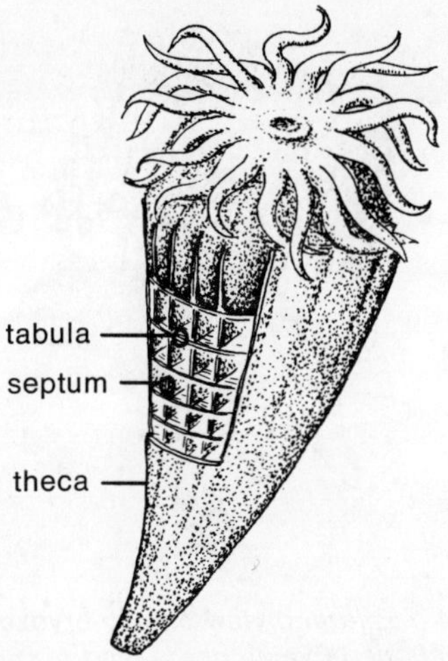

FIG. 7.11 *Coral polyp in its theca.*

include many colonial forms like the "honeycomb" and "chain" corals.

The tetracorals and tabulate corals became extinct at the end of the Paleozoic and were followed during the Mesozoic and Cenozoic by anthozoans called hexacorals as a reminder of their sixfold septal arrangement. The replacement of earlier forms by the hexacorals near the Paleozoic-Mesozoic boundary may signify that hexacorals evolved from the older tetracorals. However, it is difficult to ascertain precisely how the change in symmetry and order of septal insertion could have occurred between these two groups of coelenterates. Furthermore, hexacorals are composed of aragonite, a form of calcium carbonate which differs crystallographically from the more common calcite of the tetracorals. It may be that hexacorals arose from soft-bodied ancestors that were simply not fossilized.

Bryozoa

Bryozoans are animals that grow in colonies which may appear mosslike to the unaided eye, and which bear a superficial resemblance to certain coelenterates. The similarity, however, cannot be pressed too far, for a bryozoan has both mouth and anus at either end of a definite digestive tract (fig. 7.12). Most fossil bryozoans secrete calcareous colonies called zoaria which may grow in cylindrical, frondose, or matlike shapes. Each tiny individual is encased in a zooecium—the counterpart of the theca in corals. The zooecia appear as pinpoint depressions

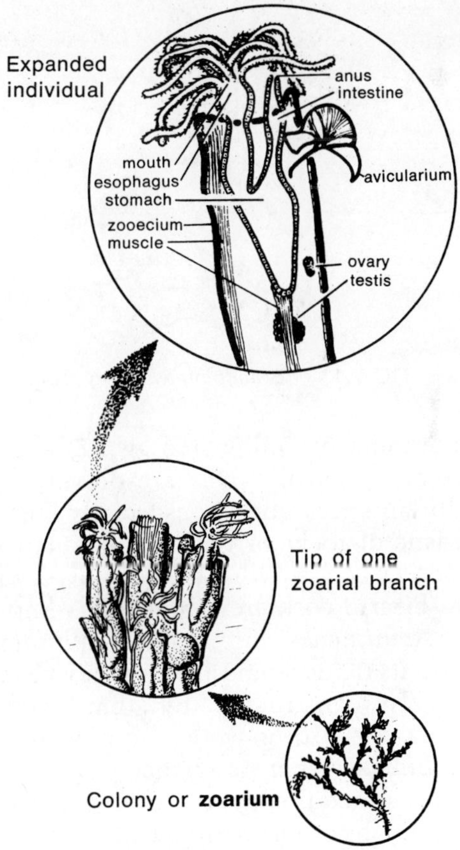

FIG. 7.12 *Bryozoa.*

on the outside of the zoaria. In most living species, the mouth is surrounded by a crescent-shaped structure bearing ciliated tentacles called the lophophore.

There are nearly 4,000 species of bryozoans living today, and nearly four times as many extinct species are known as fossils. Paleozoic colonies resembling branching twigs (fig. 7.13) are common in

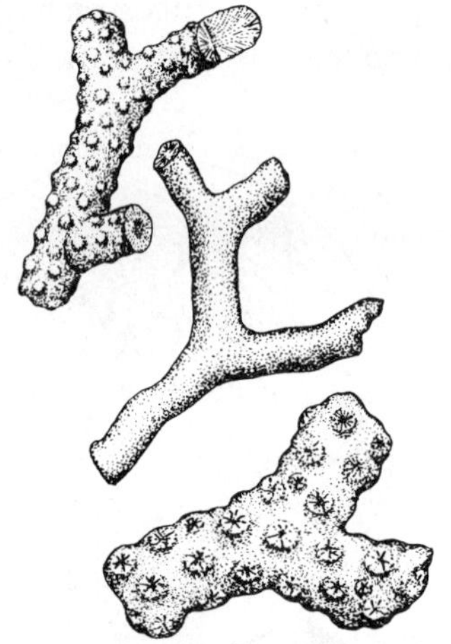

FIG. 7.13 *"Branching twig" bryozoa.*

limestones of Ordovician age. Zoaria resembling petrified lace are prevalent in Silurian strata and are especially characteristic of rocks of the Mississippian Period. Associated with the lacy bryozoans are bizarre corkscrew bryozoans known as *Archimedes* (fig. 7.14). *Archimedes* takes its name from the spiral device supposedly constructed by that famous Greek for lifting water from wells. In *Archimedes,* there is a helicoid spiral that forms an axial support. A delicate frond of lacy bryozoa is attached along the edges of the spiral axis. In life, the entire assemblage was balanced on its tip and supported by a system of oblique props that served somewhat like guy wires. Ordinarily, fossils are found with the fragile fronds broken away and only the screwlike axis remaining. *Archimedes* seems to have experienced a slow westward migration, for it is found in Mississippian strata in eastern and central North America, in Pennsylvanian rocks of our western states, and the last representatives of the genus occur in Russian sediments of Permian age. Yet another unusual fossil bryozoan is *Evactinopora* (fig. 7.14). The zoaria of this Mississippian fossil are composed of three or more vertical double layers, so that when viewed from above they resemble small petrified "stars."

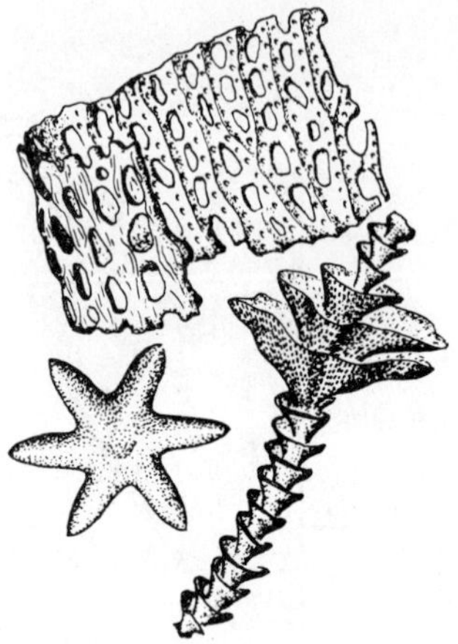

FIG. 7.14 *Enlarged view of lacy bryozoa (above), the star-shaped* Evactinopora, *and* Archimedes.

Brachiopoda

Among the most abundant, diverse, and useful fossils in Paleozoic rocks are the brachiopods. Like clams, brachiopods secrete a pair of enclosing valves (the shell), but the symmetry of the valves and the internal anatomy is different. Brachiopod shells (fig. 7.15) are almost always symmetrical on either side of the midline, and the two valves differ from

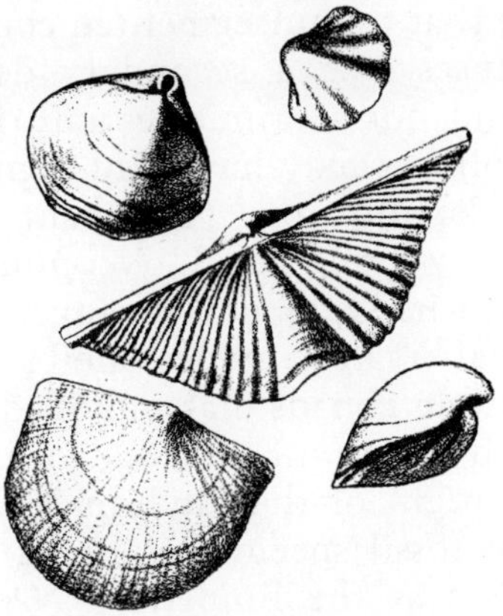

FIG. 7.15 *Brachiopods.*

one another in size and shape. The valves of pelecypod are right and left, while those of a brachiopod represent dorsal and ventral. The brachiopod shell is composed of calcium carbonate or, less commonly, mixtures of chitin and calcium phosphate. The valves may be variously ornamented by concentric growth lines, radial ridges or grooves, or spines and nodes. Some brachiopods developed a midradial fold on one valve, and a depression, or sulcus, in a corresponding position on the opposite valve. In the articulate brachiopods, the valves are hinged (they articulate) along the posterior edge and are prevented from slipping laterally by teeth and sockets. Inarticulate brachiopods lack this definite hingement, and the valves are held together only by muscles.

Among the more conspicuous soft organs of brachiopods is the lophophore. This structure consists of two ciliated coiled tentacles whose function is to keep the water between the valves in circulation and thereby more evenly distribute oxygen and remove carbon dioxide. The water currents generated by the cilia on the lophophore move food particles toward the mouth, through which these bits of food are passed into the digestive tract.

Living brachiopods are entirely marine; however, they are no longer abundant and now occupy a subordinate position among marine invertebrates. The brachiopods known to shell collectors as "lamp shells" are perhaps the best known of living species. These forms, like their Paleozoic ancestors, live attached to the seafloor by a fleshy stalk, or pedicle. More than 15,000 species of fossil brachiopods have been described. The distinctive appearance of many of these has resulted in their extensive use in correlation of strata around the world.

Mollusca

Whenever we hear the word "shellfish," we spontaneously think of members of the Phylum Mollusca, for these are the most familiar of marine invertebrates. Although most molluscans do possess shells, many such as the slugs and octopods, do not. In size, molluscans range from tiny, almost microscopic clams and snails to that largest of living invertebrates, the twenty-meter long giant squid. As presently classified, the Phylum Mollusca consists of eight classes. Three of these are termed placophorans, or "plate bearers." Placophorans are relatively primitive molluscs with multiple, paired gills. Like the snails, they move about on a muscular organ termed the "foot." Among the placophorans are the Class Monoplacophora, which have single, cap-shaped shells; the Class Polyplacophora, represented by chitons with eight overlapping shell segments; and the Class Aplacophora which lack shells altogether. The latter have no fossil record. The remaining classes of the Mollusca include the Scaphopoda, Hyolitha (a group of small molluscs with conoidal shells closed by opercula), Gastropoda (fig. 7.16), Pelecypoda (more correctly but less familiarly known as Bivalvia), and Cephalopoda. There are probably about 110,000 living species of molluscs, more than twice the number of living species of vertebrates. The molluscan name is derived from *mollis,* Latin for soft, referring to the soft body within the calcareous shell.

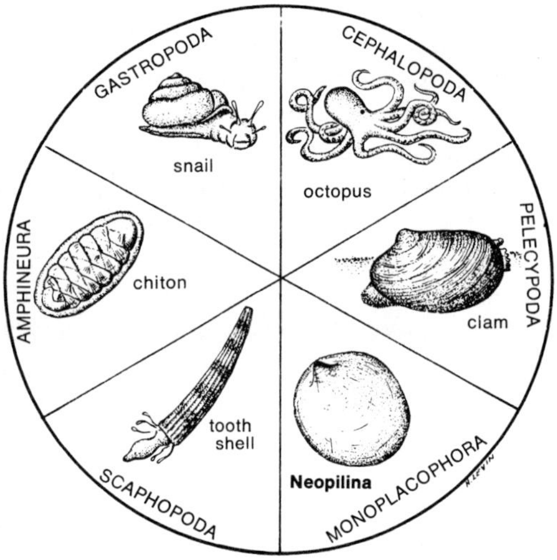

FIG. 7.16 *A diversity of molluscans.*

Molluscs are unsegmented animals and are therefore considered to have branched off the family tree of life at a point below the branch representing annelid worms. Nevertheless, the molluscs are related to the annelids. The early embryos and free-swimming larvae of annelids and molluscs are nearly identical. In addition, the mesoderm arises from a corresponding cell in both groups. Both possess nerve cords on their ventral sides, and a major dorsal blood vessel. It is quite possible that the unsegmented condition in molluscs was a secondary development, and not a primitive trait. In this way, molluscs may have had segmented annelidlike ancestors. The recent discovery of the "living fossil" *Neopilina* (fig. 7.16) with its segmental arrangement of gills, renal organs, auricles, shell muscles, and possible gonads, has provided biologists with important evidence for the segmental origin of the Phylum Mollusca. Shells of fossil species belonging to the same class as the limpetlike *Neopilina* are classified within the Monoplacophora. Fossils of monoplacophorans have been found in rocks of the Lower Cambrian and may be representatives of molluscs with primitive segmental characteristics that preceded the later unsegmented molluscans. Some biologists believe that segmentation in annelids represents an adaptation for burrowing. Thus as early molluscs developed the crawling-gliding type of locomotion, segmentation was lost.

In molluscs, a fleshy fold called the mantle has the function of secreting the shell, and a muscular portion of the body (the foot) has a primarily locomotor function. In some forms, like the squid and cuttlefish, the foot is modified into tentacles. Respiration is accomplished in aquatic molluscs by means of gills. Well-developed organs for digestion, sensation, and circulation are evidence for the advanced stage of evolution reached by this important phylum.

Pelecypoda (Bivalvia) are molluscs with a hatchet-shaped foot, layered gills, right and left valves, bilobed mantle, and the absence of a definite head. These bivalves are known by such common names as oysters, clams, and mussels. Pelecypods made their first appearance in

Cambrian time, but did not become notably diverse or numerous until the Mesozoic. Many thick-shelled oysterlike forms such as *Gryphaea* (fig. 7.17) are found in

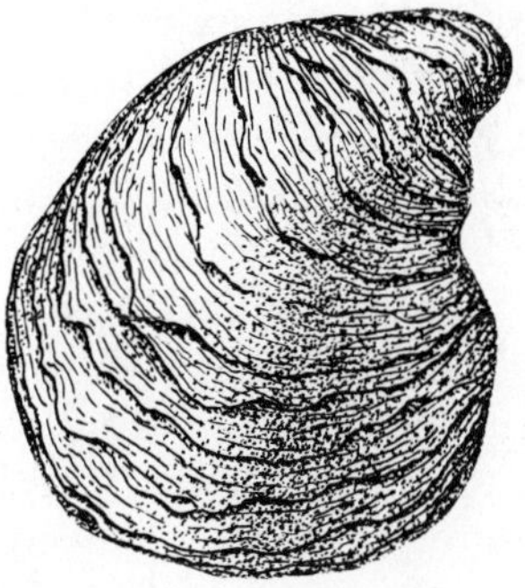

FIG. 7.17 Grypaea.

Jurassic and Cretaceous strata. Species of scallops like *Pecten* (fig. 7.18) occur frequently as fossils in Cenozoic rocks and are useful in stratigraphic correlation.

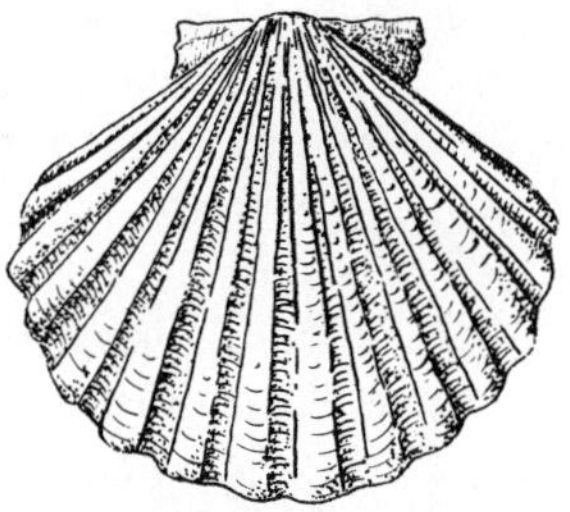

FIG. 7.18 Pecten.

The gastropods are a widely distributed and diverse class of molluscs. More than 35,000 living species and over 14,000 fossil species are known. Gastropods may lack a shell, but more commonly they construct a one-piece coiled conch. The typical gastropod has a distinct head, mouth, eyes, tentacles, and a ventrally flattened foot (fig. 7.19). They are known as fossils in the rocks of all eras after the Precambrian, and varieties have inhabited fresh water, seawater, and terrestrial

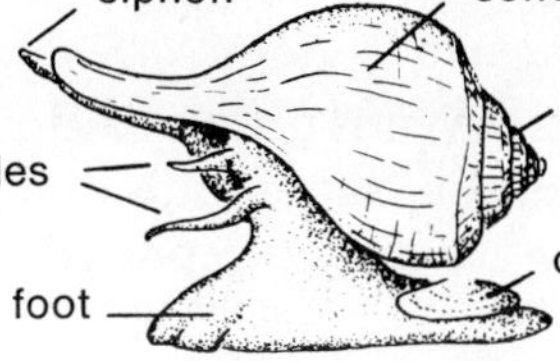

FIG. 7.19 *A gastropod.*

environments. Most gastropods are herbivorous, some are carnivorous, and a few parasitic. Like the pelecypods, their fossil remains are particularly useful in correlation of Cenozoic strata.

The cephalopods are perhaps the most complex of the molluscans. Octopods, squids, cuttlefish, and nautiloids are living examples of cephalopods. The chambered nautilus still lives and is the source of important information about the habits and soft anatomy of a vast array of fossil forms known only from their preserved conchs. Cephalopods are generally fast moving, "brainy" animals with complicated patterns of innate behavior. All are marine, and most are predaceous carnivores. Today cephalopods are not as ecologically successful as are the gastropods and pelecypods, and they probably do not exceed 350 species. However, in the geologic past, cephalopods were among the most abundant of marine animals. There are more than 7,000 fossil species known. In the genus *Nautilus,* the bilaterally symmetrical body has a prominent head which bears paired, image-forming eyes (fig. 7.20). Tentacles are present as modifications of the anterior portion of the foot. Locomotion at relatively high speeds is by jet propulsion, as water is forcefully expelled through a tubular funnel.

Superficially, conchs of cephalopods remind one of some gastropod conchs. Most cephalopods, however, coil in a plane (planispirally), whereas the majority of gastropods have helicoid conchs. An even more important difference is that the cephalopod conch is divided into chambers by transverse partitions. The

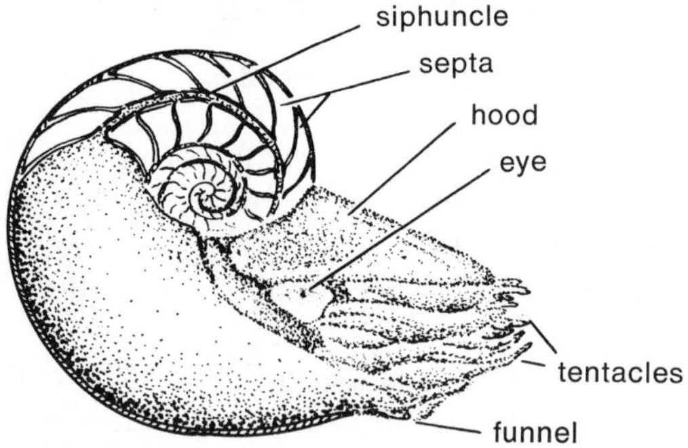

FIG. 7.20 A *cephalopod* (Nautilus *with half of shell removed*).

bulk of soft parts reside in the final chamber, and new partitions (septa) are added behind the visceral mass as the animal grows. Lines formed on the inside of the conch where the partitions join the inner wall are called sutures. The configurations of the sutures (fig. 7.21) are used for

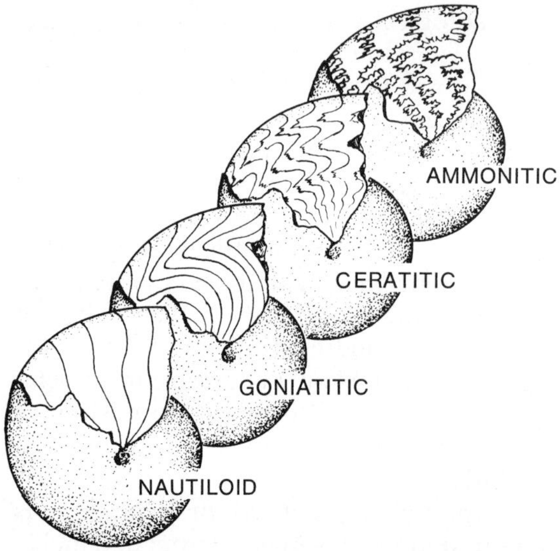

FIG. 7.21 *Cephalopod suture patterns.*

identification and classification of cephalopods. In the two great orders of shelled cephalopods, the Nautiloidea have smooth or broadly undulating sutures, and the Ammonoidea develop marked curvatures and complex wrinkles. Nautiloids were abundant and widely distributed during the Paleozoic Era, after which they declined markedly. Today the chambered nautilus is the only surviving nautiloid. Ammonoids are ideal index fossils for Mesozoic strata. Individual ammonoid lineages evolved rapidly and split into large numbers of distinct lines. As pelagic swimmers with wide ecologic tolerance, they became widely distributed throughout the marine environment. Many specializations in conch architecture, ornamentation, and suture patterns developed before the entire order suffered extinction at the end of the Cretaceous.

Some members of the Class Cephalopoda, like the octopus, lack calcareous conchs. Others have internal rather than external conchs. An example is the cuttlefish *Sepia* whose internal cuttlebone is placed in bird cages as a source of calcium for pet birds. Only one order of internally shelled cephalopods is important paleontologically. This order is the Belemnoidea. The conch of a typical belemnite resembles a cigar in size and shape. Such fossils are useful in paleontological studies of Jurassic and Cretaceous strata.

Three classes of the Phylum Mollusca are of lesser importance to geologists. These are the Monoplacophora, the Amphineura, and the Scaphopoda. Monoplacophorans, like *Neopilina* mentioned earlier, have broadly conical shells. Amphineurans (fig. 7.22) include chitons or "sea mice." Their shell is composed of eight transverse plates which cover and

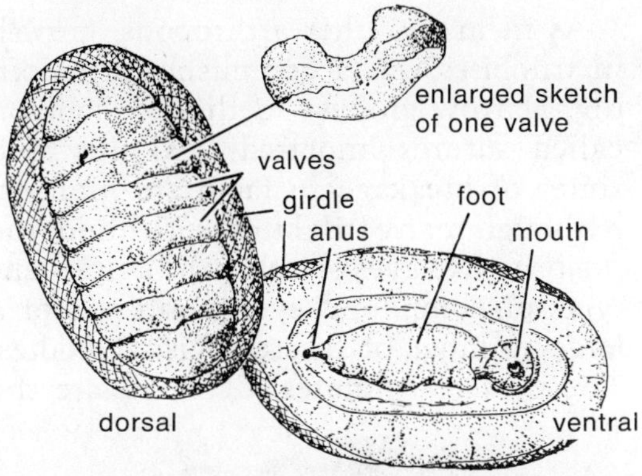

FIG. 7.22 *Amphineuran.*

protect the dorsal surface of the animal. Scaphopods or "tooth shells" are small molluscs characterized by elongate tubular shells that are open at both ends and shaped like miniature tusks (fig. 7.23).

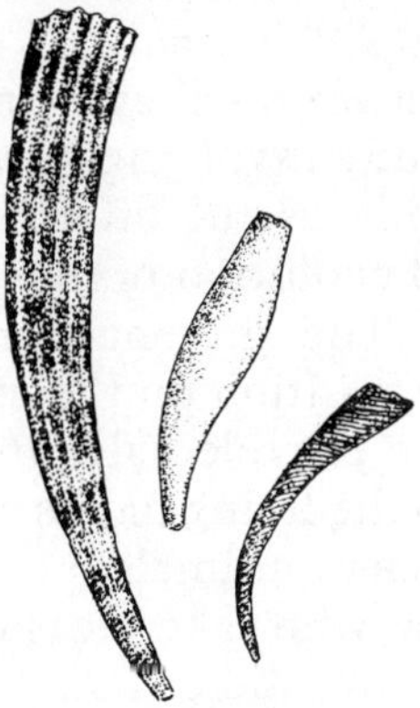

FIG. 7.23 *Scaphopods.*

Arthropoda

The number of species in the arthropod phylum is truly immense. Over 80 percent of all living animals are arthropods, and the number of individuals is astronomical. Included here are lobsters, spiders, insects, centipedes, and a host of creatures too numerous to mention. The arthropods possess chitinous exterior skeletons, segmented bodies, paired and jointed appendages, and highly specialized nerve and sensory organs. In their evolution, arthropods, along with the annelids, arose from a segmented ancestor which probably had paired appendages on each segment, and a nervous system which encircled the anterior end of the digestive tract and passed backward along the ventral surface as a double cord interrupted by segmental ganglia. These general characteristics became modified in the diverging arthropod lineages so as to provide specialized segments and appendages, division of the body into distinct functional regions in which two or more segments coalesced, and various other specializations of organ systems.

The success of arthropods in populating diverse environments is extraordinary, and yet only a relatively few groups have left a significant fossil record. The marine trilobites and ostracods (fig. 7.24) are notable exceptions and have become the most useful of all the arthropods for paleontologic studies. In terms of numbers, complexity of structure, and variability of form, the trilobites must be considered remarkably successful. They first appeared about 600 million years ago, and their abundant remains provide geologists with an arbitrary index for recognizing the beginning of the Paleozoic. Trilobites prospered all during the Early Paleozoic Era and then began a gradual decline to final and complete extinction by the end of the Permian Period.

The trilobite exoskeleton consists of a dorsal shield and a thin ventral membrane which, along with the appendages, is rarely preserved. In life, the chitin of the skeleton was strengthened by lime carbonate in areas not requiring flexibility. The term trilobite means "three lobes" and refers to the division of the dorsal shield into three longitudinal segments; namely, a central axial lobe and two lateral lobes. More readily noticed are the transverse differentiation of the dorsal shield into an anterior cephalon, a segmented thorax, and a posterior pygidium. Many of the oldest trilobites lacked distinct pygidia. This trait, along with their long series of thoracic segments is considered additional evidence that these arthropods had annelid ancestry. According to this concept, the pygidium evolved by fusion of posterior thoracic segments.

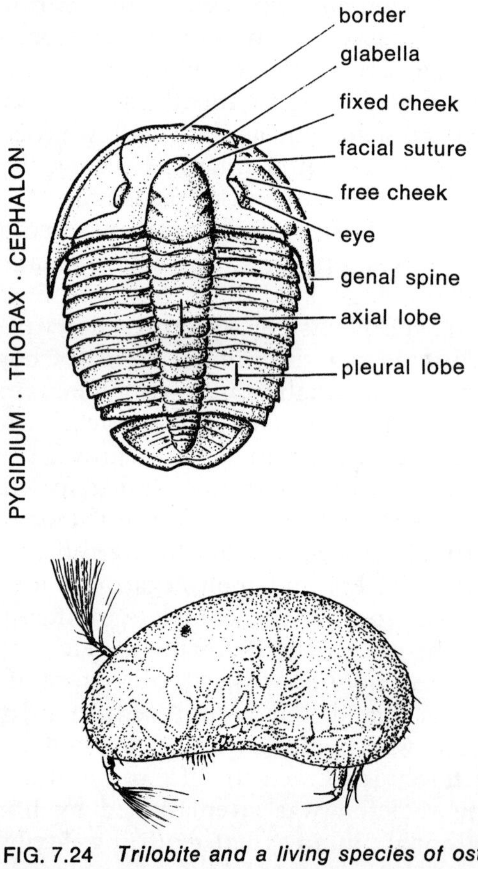

FIG. 7.24 *Trilobite and a living species of ostracode.*

As in many other arthropods, growth in trilobites was accomplished by molting. Narrow grooves in the exoskeleton called sutures imparted flexibility and zones of breakage to facilitate molting. Although many trilobites were blind, the majority had eyes of either the single-lens type, or compound eyes composed of a large number of discrete visual bodies. The eyes of trilobites (fig. 7.25) are the

FIG. 7.25 *Side view of the trilobite* Phacops rana *showing the discrete visual bodies of the eye.*

earliest known fossil eyes and, with continuing study involving new techniques, may provide useful information on the origin and evolution of invertebrate eyes generally. The presence or absence of eyes, their position on the cephalon, and their shape provide paleontologists with clues as to the living habits of these fascinating ancient animals.

To those who take pleasure in reflecting on nature, there is an air of mystery and sadness that comes with the discovery of a trilobite in a ledge of rock along a rarely visited creek bed. T. A. Conrad captured that feeling when he wrote:

> The race of man shall perish, but the eyes
> of trilobites eternal be in stone,
> And seem to stare about in mild surprise
> At changes greater than they have yet known.

The second group of paleontologically important arthropods is the ostracods. At first glance, ostracods might appear so different from trilobites as to cause one to doubt their classification within the same taxonomic class. Small, bean-shaped crustaceans, the ostracods have a bivalved shell vaguely suggestive of some sort of pelecypod. However, the shell, or carapace, encloses a segmented body from which extend seven pairs of appendages. The adult ostracods are about half a millimeter to as much as four millimeters in length. The two valves of the carapace are composed of chitin and calcium carbonate and are hinged along the dorsal margin. Ostracods first appeared in the Early Ordovician and continue to inhabit seas and freshwater bodies to the present day.

Echinodermata

Echinoderms (fig. 7.26) are highly developed animals, typically with pentamerous adult symmetry and a knobby or plated skeleton. A unique characteristic of the phylum is the presence of a tubular water vascular system which functions in respiration and locomotion. Among the many classes of the Phylum Echinodermata, the Asteroidea (starfish), Ophiuroidea (brittle stars), Echinoidea (sea urchins), Crinoidea (sea lilies), Blastoidea, and Cystoidea are most abundant and useful in stratigraphic studies.

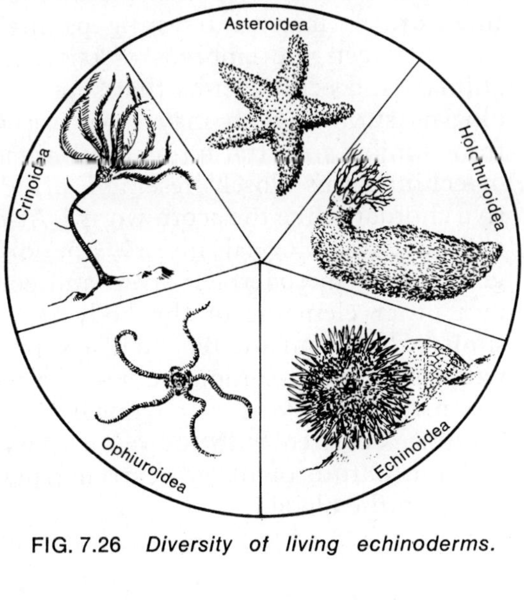

FIG. 7.26 *Diversity of living echinoderms.*

The asteroids are the most popular of the echinoderms. Typically, they have five arms which radiate outward from an ill-defined central disc. Portions of the water vascular, digestive, and reproductive systems are contained within the arms. Asteroids range in age from the Ordovician to the present, but are not as common in Paleozoic strata as the attached echinoderms.

The ophiuroids differ from the asteroids in having sharply delimited central discs and long serpentine arms. Unlike the starfish, the arms lack food grooves; food is taken in only at the mouth on the ventral side of the central disc. The ophiuroids began their evolutionary history along with the starfish in the Ordovician Period.

Echinoids are a class of echinoderms represented by the common sea urchin and sand dollar. Shape is variable from spheroidal to discoidal. Five grooves with openings for the tiny tube feet of the water vascular system are typically present. Commonly in the spheroidal forms, the mouth is located on the ventral surface and the anus on the opposite or upper surface. However, in many discoidal echinoids, the anus has migrated to the posterior margin. Although echinoids have been found in rocks as old as Ordovician, they did not become plentiful until the middle part of the Mesozoic Era.

The crinoids, blastoids, and cystoids are the stemmed, or attached, members of the Phylum Echinodermata. Crinoids are composed of three main parts; namely, the calyx, the arms, and the stem (fig. 7.27). Most of the vital organs are encased in the calyx. The arms bear ciliated food grooves and function in bringing food particles down toward the mouth. Disaggregated plates of crinoids are found in rocks which range in age from Ordovician to the Recent. They were most numerous and varied during the Mississippian Period. The blastoids had a shorter time range, suffering total extinction in the Permian. These interesting fossils had threadlike filaments which extended from the five food-groove areas of the budlike calyx, and functioned much as do the arms of crinoids.

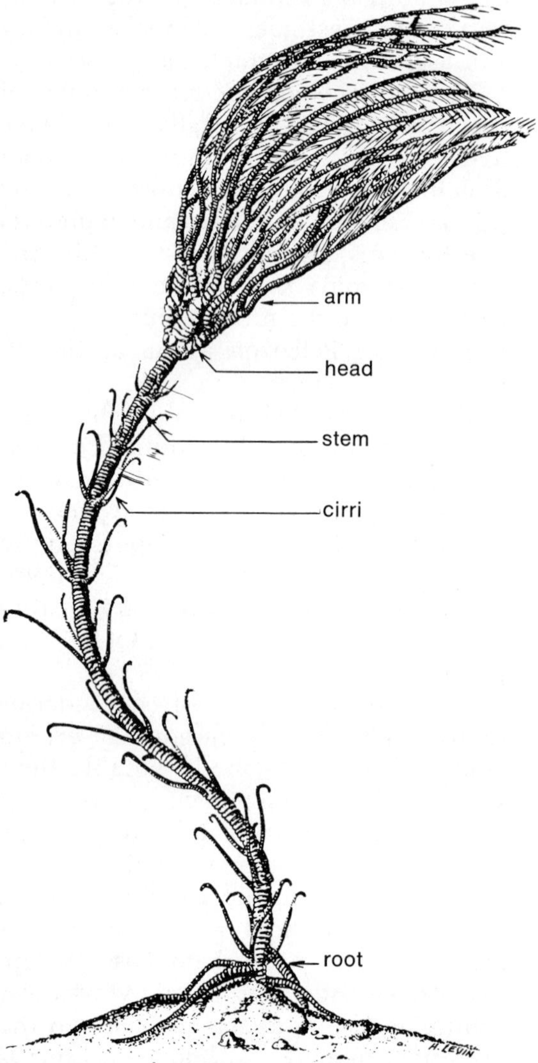

FIG. 7.27 *Crinoid.*

The cystoids are a more primitive group than the blastoids and crinoids. The pentamerous symmetry which characterizes most echinoderms is poorly shown or even absent in cystoids. The plates of the calyx may be either symmetrically or asymmetrically arranged and bear pores. Although their complete geologic range is from Cambrian to Late Devonian, cystoids are found chiefly in Ordovician and Silurian rocks.

Echinoderms, with their radial symmetry, lack of internal skeleton, and distinctive water vascular system, seem unlikely candidates for ancestors or relatives of vertebrate animals. However, there are so many interesting parallelisms between the embryos of chordate animals and echinoderms that some zoologists surmise both may have arisen from similar ancestral forms. The larvae of echinoderms closely resemble primitive chordates like the acorn worms. Also in the embryological history, the origin of the mesodermal layer and certain other elements of the body show similar development. Biochemistry provides intriguing additional clues to a relationship by revealing chemical similarities associated with muscle activity, and the nature of oxygen-carrying pigments in the blood.

RELATIONSHIPS AMONG INVERTEBRATES

Having completed our very general survey of the fossil phyla of invertebrates, we may pause to reflect upon the general course of their evolution. One way to do this is to compare the various groups according to their general body plan and make inferences about their possible ancestral relationships. Stated simply, this method uses the assumption that the more closely the body plans of two phyla resemble one another, the closer their relationship and the more likely they had a common ancestry. Often it is the premature stages that are the most useful, for adults tend to mask fundamental similarities behind recently acquired adaptations. In a broad sense, animals of a particular phylum tend to look most alike in their early embryonic stages, and then become gradually differentiated later as structures characteristic of the various subdivisions of the taxonomic group are produced. Carried a step further, one notices that—provided we restrict ourselves to the examination of very early embryonic stages—even different phyla look similar. Such observations led the biologist Haeckel to formulate a "principle of recapitulation" which stipulated that every animal in its individual development from egg to adult passes through a series of steps that correspond to the long evolutionary history of the larger group. Today we know that Haeckel's concept is not valid except as a broad generalization with definite limitations. In scores of animals, the embryos as well as the adults undergo evolution, thus altering the sequence. In the earliest stages of development, some changes are the inevitable result of the mechanics of changing from a single-celled fertilized egg to a many-layered multicellular form, and these changes cannot be accomplished except in one general way. Recapitulation contributes to our understanding of the relationships of animals, however. The fertilized egg of higher invertebrates shows traits comparable to those found in a unicelled protozoan. The egg divides repeatedly to form a hollow ball of cells, the blastula, which reminds one of colonies of unicellular organisms like *Volvox* (fig. 7.28). By pro-

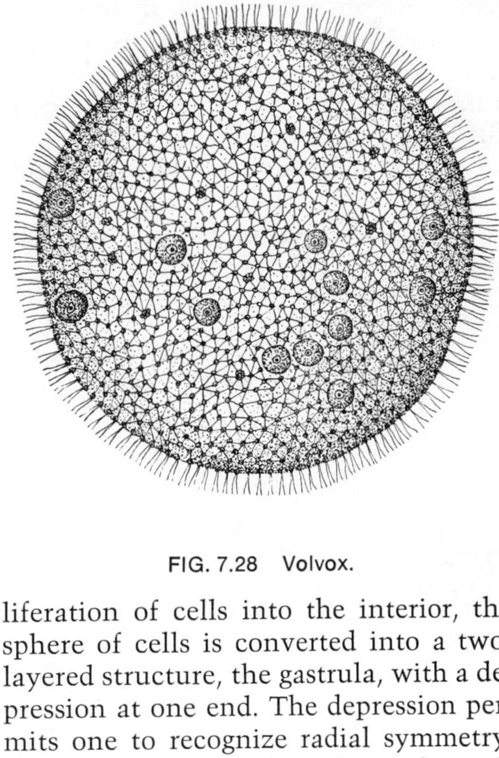

FIG. 7.28 Volvox.

liferation of cells into the interior, the sphere of cells is converted into a two-layered structure, the gastrula, with a depression at one end. The depression permits one to recognize radial symmetry, and, with the layered structure, the coelenterate stage of evolution is approximated. Coelenterates, with their ciliated ectoderm specialized for locomotion, protection, and sensation approximate this tissue level of organization. The next step was very likely the development of a mesoblast. The mesoblast is a tissue between the outer layer of cells (epiblast) and the inner layer (hypoblast). Tissues from the epiblast are called ectoderm, those from the mesoblast are mesoderm, and those from the hypoblast, endoderm. From the mesoblast, more definite organs and organ systems may originate and pave the way for the development of larger and more complex organisms than were formerly possible.

There are two ways by which mesoderm develops in animals. In molluscs, arthropods, and certain worms, it originates from two special cells in the gastrula called primitive mesodermal cells (fig. 7.29). As mentioned earlier, chordates and echinoderms have a similar manner of mesoderm origin; namely, from outpocketings of the primitive endoderm. This difference in mesodermal origin permits us to recognize two branches of the family tree of animals, the arthropod and chordate branches. The lower animals in the arthropod branch show many similar developmental traits. Except for the annelids, they are unsegmented and have very similar larval stages (trochophore larvae). With the appearance of segmentation in the annelids, arthropods arose and expanded into the groups familiar to us today. The Cambrian trilobites, many of which had a great length of body segments and lacked definite pygidia, seem to provide evidence of this annelid ancestry.

On the other branch of the family tree we find two prominent phyla, the chordates and the echinoderms, linked by the developmental similarities previously mentioned. It seems probable that the ancestral chordate, from which the vertebrates descended, may have been a relatively slow moving, or even attached, filter-feeding organism. It would have a

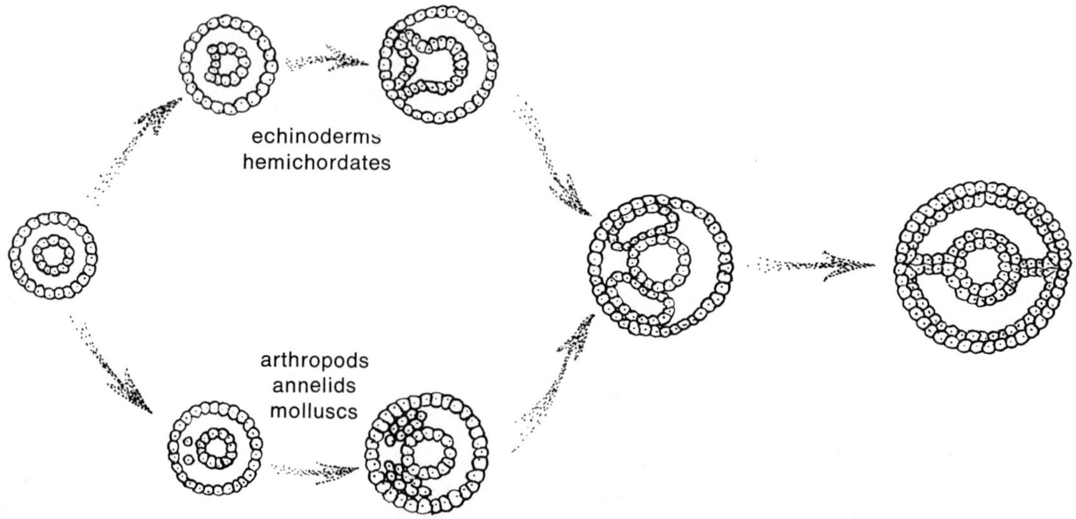

FIG. 7.29 *Schematic sections of an embryo beginning with the gastrula stage, illustrating two methods by which the mesoderm and coelum originate. (Modified from J. M. Weller, 1969,* The Course of Evolution, *McGraw-Hill, p. 93.)*

dipleurula type of larva (fig. 7.30) such as is found in primitive chordates and in echinoderms, and would thus differ from the arthropod line with its dissimilar trochophorid larval stage.

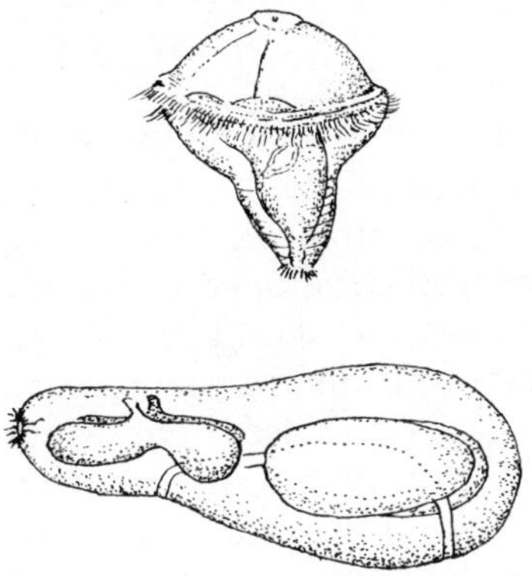

FIG. 7.30 *Trochophore and dipleurula larvae.*

SUMMARY

Early indications of life in the Precambrian consist largely of structures formed by algae, bacteria, and fungi. Unquestioned animal remains appear only in very late Precambrian strata such as those exposed at Ediacara Hills, Australia. These fossils of rather complex animals were probably preceded by a lengthy period of evolution of metazoans for which we have scant evidence. However, the fossil record improves spectacularly at the beginning of Cambrian time. Cambrian strata yield the first abundant and complex remains of invertebrates. Indeed, nearly all major invertebrate phyla with preservable hard parts are found at or near the Precambrian-Cambrian boundary. Eight phyla with mineralized skeletons that occur abundantly as fossils in Paleozoic and younger rocks are the Sarcodina, Porifera, Coelenterata, Bryozoa, Brachiopoda, Mollusca, Arthropoda, and Echinodermata. Members of these phyla evolved gradually, changing from geologic period to geologic period, so that the rocks of each period contain distinctive fossil assemblages. These fossil assemblages are used as indicators of particular units in the geologic time scale. At certain times in the history of the earth, environmental change resulted in widespread extinctions of animal life. These interruptions serve as markers to the boundaries of the major units on the geologic time scale. They were usually followed by rapid evolutionary radiations as animals expanded into the vacated ecological niches.

Among the unicelled sarcodinids, the principal groups commonly fossilized are the radiolaria and foraminifers. The Order Foraminiferida is composed of microscopic animals that secrete a usually calcareous ($CaCO_3$) test composed of one or more variously arranged and shaped chambers. The "forams" first appear in Ordovician rocks. In correlating Late Paleozoic strata, the forams with the shapes of grains of wheat, the fusulinids, are particularly useful because of their abundance and diversity. During the Cretaceous Period, pelagic foraminifers make their appearance. In general, the tests of forams are more frequently encountered in consolidated sedimentary rocks than radiolarians. The fragile, delicate siliceous tests of radiolarians are more easily crushed or dissolved after burial.

The sponges (Porifera) represent the lowest grade of multicellular animals. They lack true tissues and organs, but do have skeletal parts called spicules composed of calcite, silica, or spongin. Fossils of sponges are present in Cambrian rocks, and there are questionable discoveries even in sediments of the Late Precambrian. Sponges are classified primarily on the basis of the composition and shape of their spicules.

The Archaeocyatha was a phylum of spongelike creatures that flourished for a time in the Cambrian Period. The skeleton was vase-shaped and composed of a double calcareous wall separated by septa. The soft parts are unknown.

Coelenterates have polyp or medusoid shapes, radial symmetry, and body walls composed of two layers of cells with jellylike material between them. There is a saclike digestive cavity in coelenterates, with a single opening or "mouth." Tentacles bearing stinging cells are distinctive features of this almost entirely marine phylum of animals. Earliest coelenterate fossils are derived from rocks of Late Precambrian age. Because they are readily fossilized, stony corals of the Class Anthozoa have always been the most important coelenterates for paleontological study. Many Paleozoic tabulate and tetracorals (rugose corals) are useful guide fossils, whereas in post-Paleozoic rocks, scleractinian corals are the most useful paleontologically. The occurrence of fossil coral reefs in ancient rocks strongly implies a former shallow marine environment with relatively warm water.

Bryozoans are aquatic invertebrates that build a calcareous framework within which live tiny, polyplike individual animals. They are known as fossils from rocks of the Ordovician (questionably Late Cambrian) onward.

Brachiopods are common fossils in Lower Cambrian rocks and continue to be abundant and diverse throughout the Paleozoic Era. Brachiopods are bivalves, with one valve usually slightly larger than the other, and both valves bilaterally symmetrical. They are marine animals which attach themselves to the seafloor by a fleshy pedicle and filter feed with the aid of ciliated tentacles supported on a lophophore.

Members of the Phylum Mollusca are known as fossils in rocks as old as Cambrian. Many develop shells of calcium carbonate that is secreted by a fold of tissue (the mantle) which covers the body. Digestive, circulatory, and nervous systems are well developed in the molluscs. The diversity of animals in this great phylum is reflected in the characteristics of its six major classes. The Monoplacophora are primitive molluscs with single, cap-shaped shells, whereas a "shell" of eight overlapping segments characterizes the Amphineura. The most abundant molluscs are Gastropoda which commonly have shells coiled spirally. The Pelecypoda (Bivalvia) have a flattened body between mirror-image right and left valves. The Scaphopoda are the "tusk-shelled" molluscs. Finally, the Cephalopoda are a marine class either with conchs (as in the chambered nautilus) or without (as in the octopus). Conchs of fossil cephalopods may be straight or coiled, chambered, and characterized by smooth or variously fluted septa that separate the chambers.

In general, animals classified as Arthropoda have chitinous external skeletons, jointed appendages, and segmented bodies. The phylum includes insects, millipedes, centipedes, crustaceans like lobsters and ostracodes, and spiders. An especially important fossil class of arthropods is the Class Trilobita. The first occurrence of trilobites 600 million years ago marks the beginning of the Paleozoic Era. The trilobites have a segmented, longitudinally trilobed exoskeleton. Trilobites did not survive beyond the Permian.

Echinoderms are exclusively marine animals that have been common as fossils since the Cambrian. Radial symmetry is characteristic of adult echinoderms, although the larvae are bilaterally symmetrical. Using a system of water-filled tubes comprising a water vascular system, echinoderms are able to catch food and move about on the seafloors. Among the many classes of this phylum are the free-living Asteroidea (starfish), Echinoidea (sea urchins), Ophiuroidea (brittle stars); and the attached Crinoidea, Blastoidea, and Cystoidea. In addition to the above classes, there are several classes of lesser paleontologic importance.

A question of some evolutionary interest is the relationship of the invertebrate classes to one another. Because chordates and echinoderms have a similar manner of mesodermal origin in the developing embryo and similar forms of dipleurula larvae as well, it is generally believed that both groups sprang from a common ancestor. In contrast, arthropods, and also annelids, brachiopods, molluscs, and ectoproct bryozoans have the embryonic mesoderm arising from two special primitive mesodermal cells in a manner distinctly different than that seen in echinoderms. The difference in this so-called arthropod line is also apparent in the different form of the free-swimming larvae.

SUPPLEMENTARY READINGS AND REFERENCES

Beerbower, J. R. 1968. *Search for the past.* Englewood Cliffs, N.J.: Prentice-Hall.

Buschbaum, Ralph. 1948. *Animals without backbones.* Chicago: University of Chicago Press.

Geological Society of America. 1953–present. *Treatise on invertebrate paleontology.* Geological Society of America and University of Kansas Press.

Russell-Hunter, W. D. 1968. *A biology of the lower invertebrates.* New York: Macmillan Co.

Tasch, Paul. 1973. *Paleobiology of the invertebrates.* New York: Wiley.

Weller, J. M. 1969. *The course of evolution.* New York: McGraw-Hill.

Wilmoth, J. H. 1967. *Biology of Invertebrata,* Englewood Cliffs, N.J.: Prentice-Hall.

8 The Flexible Rod

A fish-like thing appeared among the annelids one day,
It hadn't any parapods or setae to display,
It hadn't any eyes or jaws or ventral nervous cord,
But it had a set of gill slits and it had a notochord.

—From the song, *Amphioxus*

TO BE A CHORDATE

In the flickering greens of ancient Paleozoic seas, there lived humble creatures whose descendents were destined to change the very face of the earth. The ancestors of these animals, we now call chordates. We do not know their precise appearance, but think they were sedentary animals that lived attached to the seafloor where they gathered tiny particles of food in their ciliated arms. Perhaps one day some lucky paleontologist will blow the dust from a rock ledge and discover the impression of one of these earliest chordates. However, until that time, we must be satisfied with examining living creatures that we think approximate the appearance of the earliest members of the chordate pedigree.

For full membership into the phylum chordata, an animal should possess, at least in some stage of its life history, some sort of longitudinal supportive structure (notochord or vertebral column); a dorsal, hollow central nervous system; gill slits; a postanal tail; and blood which circulates forward in the main ventral vessel and backward in the dorsal.

The notochord, from which the name chordata is derived, is clearly visible in cross sections of the frog tadpole or the adult animal known as *Branchiostoma* (fig. 8.1). In it are found large vacuolated

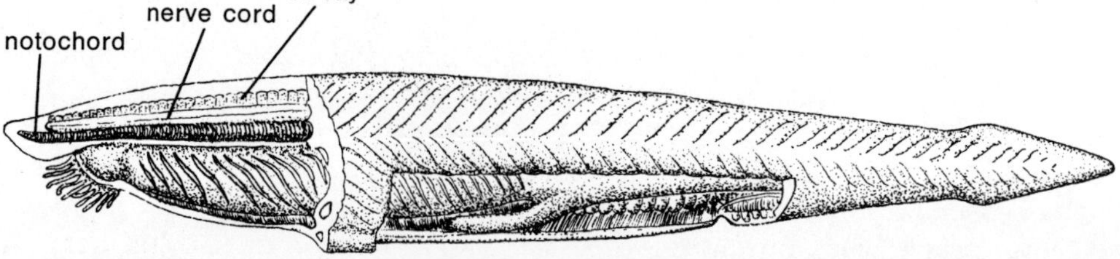

FIG. 8.1 Branchiostoma. *(Modified from Storer, Usinger, and Nybakken, 1968,* Elements of Zoology, *McGraw-Hill, p. 375.)*

cells containing fluid under pressure. Such cells give the notochord rigidity and elasticity simultaneously.

The hollow, dorsal central nervous system of chordate animals is decidedly different from the nervous systems of other metazoans. It is formed as a tube by infolding of a dorsal band of ectodermal cells in the embryo. The hollow within the tube may be much reduced, but it is always visible in chordates. By comparison, all metazoans above the level of sponges have cells specialized for conducting impulses. These cells are ordinarily concentrated into tracts of a central nervous system. Usually such tracts consist of series of knots or ganglia connected by cords and, more importantly, the entire system is generally on the ventral side of the organism.

Gill slits, yet another diagnostic feature of chordates, have no precise counterparts in other animals. Sometimes called pharangeal clefts, they are paired, segmentally arranged openings extending through the pharynx to the exterior of the animal. Water taken in at the mouth is passed through these openings. Gill slits are absent in the adults of terrestrial chordates, although they may be partially or temporarily present in embryonic stages.

As chordates without tails, we may—at least for an initial moment—feel surprise that this organ is a salient chordate feature. Yet in man, the coccyx, already vestigial in six-week-old embryos, is the remnant of a tail consisting of five poorly developed tail vertebrae. Along with man, the adult frog also lacks a tail, but the tail is obvious in the tadpole. In bilaterally symmetrical nonchordates, the anus—at least in the embryo—is located at the very end of the body. An apparent exception occurs in snails, but in this group the more forward position of the anus is the result of twisting of the body during development.

The chordate circulatory system is distinctively different from that of nonchordates. The ventral vessel drives the blood forward and is developed at one point into a pumping organ. In nonchordates which have a blood vascular system, the blood never flows forward in the primary ventral vessel, nor is that vessel ever modified into a heart.

DRAMATIS PERSONAE

The drama of vertebrate beginnings cannot be adequately viewed without pausing to describe the principal characters in the cast. They are indicated in figure 8.2.

As can be seen, the Phylum Protochordata is customarily divided into three classes of prevertebrates. The first of these is the Enteropneusta. Enteropneustans are often called acorn worms in reference to the acorn like appearance of the proboscis nestled in the encircling collar. Among the twelve genera of enteropneustans, *Balanglossus* (fig. 8.3) is perhaps the most widely known. Balanglossids are essentially sessile burrowers in the sand or mud of seashores. The proboscis is pushed forcibly into the sediment pulling the collar along with it. The distended collar anchors the anterior end of the animal, and contraction of muscles in the trunk draws the remainder of the body forward. The mouth in *Balanglossus* is ventrally located just inside the collar. Water entering the mouth is passed out through the pharangeal gill slits. Both a dorsal and ventral nerve cord are present, but only the dorsal cord extends into the

TAXA	DESCRIPTION & AGE	APPEARANCE
Phylum Protochordata	Mud straining organisms with some trace of notochord and gill slits. Non-segmented, and having a muscular collar which may bear arms. (Cambrian to Recent)	
Class Enteropneusta	Filter feeding worm like forms with simple collar and muscular proboscis. (Recent)	 acorn worm
Class Pterobranchia	Filter feeding colonial forms with chitinoid skeleton and armless collar. No sicula (Ord.-Rec.)	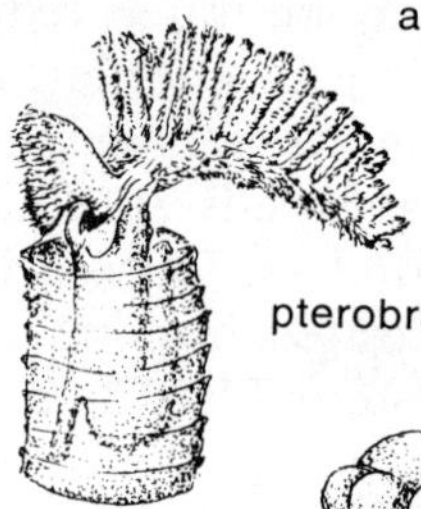 pterobranch
Class Graptolithina	Filter feeding colonial forms with chitinoid skeleton and sicula. Soft parts unknown. (Camb.-Miss.)	 graptolite colony
Phylum Chordata	Animals possessing notochords and gill slits in some stage of development. Arms not developed. Level of organization approximates that of arthropods. (Ordovician to Recent.)	
Subphylum Tunicata	Filter feeders with notochord in tail portion of larva. Adult with large gill basket enclosed in sac like tunic. (Recent.)	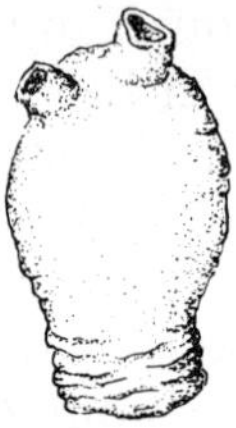 adult tunicate  larval tunicate
Subphylum Cephalochordata	Segmented filter feeders with large notochord that extends to the head. Large gill structure and fishlike shape. (Recent.)	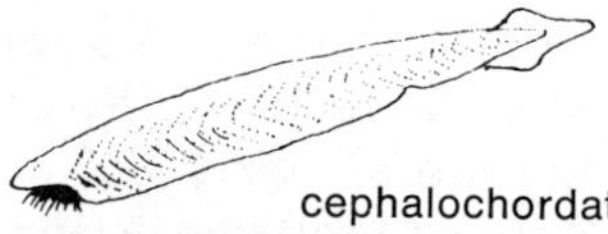 cephalochordate
Subphylum Vertebrata	Chordates in which the notochord is supplemented or replaced by a series of cartilagenous or bony vertebrae.	 vertebrate

FIG. 8.2 *General characteristics of protochordates and chordates.*

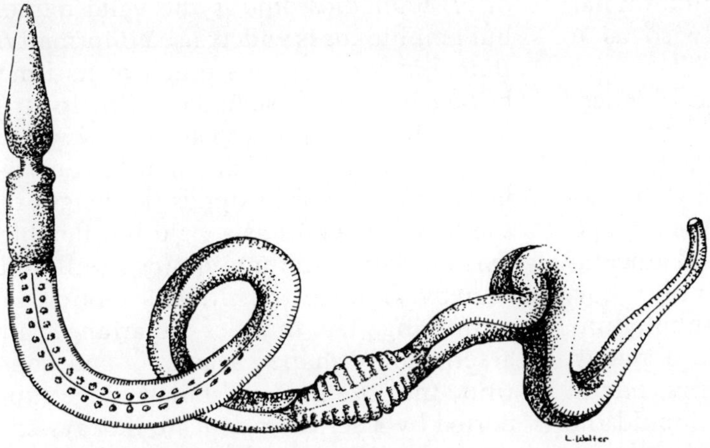

FIG. 8.3 *The "acorn worm,"* Balanglossus.

collar where it thickens and is hollow, and thus leads one to associate it with the tubular dorsal nerve cord of the true chordate. The notochord is thought to be represented by a short, rodlike outgrowth which extends from the anterior end of the digestive tract into the base of the proboscis. Like the notochord in the more advanced cephalochordates, it is composed of vacuolated cells. Relevant to the question of protochordate origins is the fact that the larva of acorn worms resemble, to a surprising degree, the larva of certain echinoderms.

The pterobranchs (fig. 8.2) are tiny colonial animals that live a sessile existence. They reside within a secreted tube and possess a proboscis, collar, and ciliated tentacles for the entrapment of food. These ciliated tentacles resemble the lophophore feeding mechanism of brachiopods and bryozoans. The simple notochord is similar to that seen in enteropneustans. A single pair of gill slits are found in two genera (*Cephalodiscus* and *Picoryne*) but none exists in the remaining genus (*Rhabdopleura*).

The third group of protochordates are the enigmatic Graptolithina (fig. 8.2 and 8.4). Because these animals have been extinct for over two hundred fifty million years, the nature of their soft tissues must be inferred. Recent studies of superb uncompressed graptolites reveal their similarities to such living pterobranchs as *Rhabdopleura* and are the basis for including these fossils in the Protochordata. Pterobranchs and graptolites lived in the enclosed tubes they secreted. Both groups were capable of reproduction by asexual budding from the parent colony. In both, the structure of the enclosing sheath con-

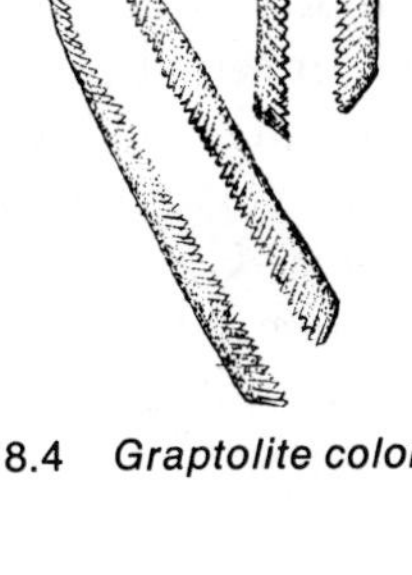

FIG. 8.4 *Graptolite colonies.*

sists of alternating bands which form half segments that meet medially so as to form a zigzag line.

A step upward to the more advanced Phylum Chordata are the Tunicata. Sea squirts, sea potatoes, sea peaches, sea pork, and sea plums are among the gastronomic names applied to these unappetizing creatures. The name "Tunicata" refers to the tough, translucent tunic covering most members of the subphylum. The simplest forms resemble a pudgy, upright vase with two openings, one at the side, and one at the top. Should the animal be frightened, the body wall may suddenly contract, so that the water contained inside is forcibly ejected. This action produces the "squirt" to which the name sea squirt refers. Within the tunic is a large saclike pharynx perforated by many rows of gill slits. Except for these gill slits, there is little about this animal to suggest its chordate pedigree. However, the tunicate larva tells us a different story. Unlike the pudgy sedentary adult, it is an active, free-swimming "tadpole" with a well-developed tail, notochord, and tubular dorsal nerve cord.

In many ways, the Cephalochordata are the most advanced of the prevertebrates. They are small fishlike animals that include about thirty species. The important chordate features of dorsal hollow nerve cord, notochord, and gill slits are present throughout the lives of these creatures. The most widely known cephalochordate is *Branchiostoma*, which many believe closely resembles the hypothetical ancestor of the vertebrates. The organism was originally named in 1934 by Costa. Without knowing of the earlier name, Yarrell named the creature "amphioxus" in 1936. Because of the rules of nomenclatural priority, *Branchiostoma* is the valid name, but amphioxus is widely used informally. *Branchiostoma* spends much of its time buried in bottom sediment with its anterior end protruding so as to draw water and food in through the mouth. A cartilagelike notochord extends the length of the body. It serves as a rigid but flexible axis on which the muscles can pull and thereby produce the sinuous motion of swimming. The muscles are arranged in a series of V-shaped segments and continue into the postanal tail. A ridge supported by a series of delicate fin rays extends along the dorsal surface and around the posterior end as a caudal fin. Lying between the dorsal surface and the notochord is a tubular nerve cord (fig. 8.1). As befitting a prevertebrate, the pharynx is perforated by gill slits which function mainly as organs for straining food from water. However, the nerve cord does not expand into an anterior brain, but tapers to a point. Many believe this is a recent development in which there was degeneration of parts of the head region in response to the sedentary habits of this particular animal. In this, and in other ways, *Branchiostoma* is very specialized. Nevertheless, it serves as a living model for obtaining an idea of the appearance of the ancestral link to the vertebrates.

TO FIND AN ANCESTOR

The ancestral forms from which vertebrates arose have been sought in several diverse groups of invertebrate animals, especially the annelids, arthropods, and echinoderms. Theories of origin favoring the annelids and arthropods were once held in high esteem. However, these theories encountered difficulty in attempting to explain the difference in position of the nerve cord which is dorsal (relative to the digestive tract) in chordates, but ventral in both arthropods and annelids. An even more serious problem concerns the totally different stages in the early development of the fertilized egg. The arthropod egg cleaves spirally (fig. 8.5), and the coelum opens in the

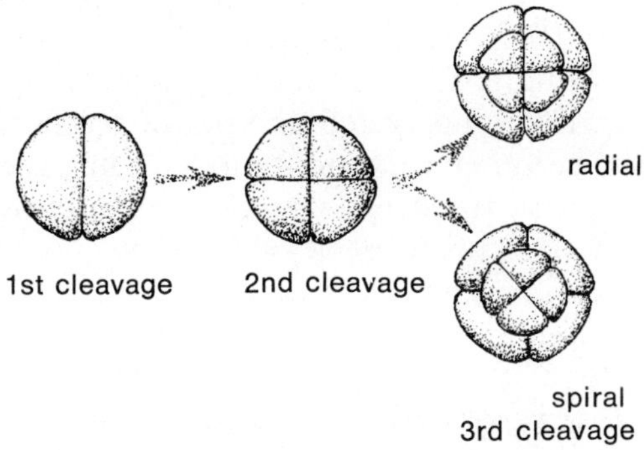

FIG. 8.5 *Comparison of radial and spiral cleavage of an egg. (Modified from J. M. Weller, 1969,* The Course of Evolution, *McGraw-Hill, p. 87.)*

midst of early mesodermal cells that move inward from the entoderm. On the other hand, in invertebrate chordates, the cleavage is radial (fig. 8.5) and the coelum originates by invaginations of the entoderm.

In the face of these arguments, most investigators favor the theory that both vertebrates and echinoderms evolved from a common ancestor. They support their belief with observations of similarity in the embryonic development of echinoderms and chordates. Both, after all, have bilaterally symmetrical dipleurula larvae. The skeletons of both groups originate mesodermally, their blood proteins are similar, and their muscle activity is dependent upon the same kind of phosphorous compound that is utilized in a manner different from that in other animals.

What was the pathway by which the vertebrates arose? Alfred Romer (1967) suggests that the steps up the evolutionary ladder to man began among arm-feeding metazoans that lived sedentary lives on the floors of the Cambrian seas. They occur abundantly as fossils and their descendents include the familiar crinoids. In regard to chordate origins, it is significant that the tiny living pterobranch member of the Phylum Protochordata is itself such a sessile arm feeder.

An engineer observing the feeding of the sessile arm bearer, might—with machine age cunning—seek to design a better way of bringing in food. "Why, the method is not much better than the way a mop catches dust particles," our toolmaker observes. "Far better if it were made more on the order of a vacuum cleaner." Without intending to slight man's creativity, it seems nature has a remarkable capacity for preconceiving the designs of inventors. The dust mop to vacuum cleaner transition may be followed in the development of sessile armless filter feeders from the sessile arm feeders. This important progression was accomplished by bringing the cilia "indoors" and by developing exhaust vents, the gill slits. Gill slits are present in simplest form in the pterobranch *Cephalodiscus.* In this protochordate, water containing food particles is drawn into the mouth and hastened through the gills by the rhythmic strokes of bands of cilia. Food is retained in mucus as the water leaves the pharynx through the gill slits. The particles are then conveyed by cilia along definite paths to the gut. Such may also have been the feeding mechanism of the descendents of the arm feeders. Indeed, with the development of the gill current, arms lost their function in feeding and were eliminated. Much later, muscle action replaced the current producing function of cilia as animals adapted themselves to a diet of larger food. The gills no longer served as food strainers and assumed their well-known function as organs for breathing.

But let us return momentarily to our engineer. His intention was certainly to produce a movable vacuum cleaner. It is one thing to suck food laden particles into an organism, but quite another to go in search of nourishment. Certainly, in periods of scarcity of food, the animal with the capacity for exploration has an advantage over his sedentary neighbors. We have speculated about the evolution of sessile arm feeders to armless sessile gill feeders like the adult tunicate. Vagrant filter feeders were an almost inevitable next development.

The precise manner in which this new evolutionary development occurred cannot be ascertained paleontologically, for the fossil evidence is lacking. Biology, and particularly the concept of larval evolution, does provide a plausible explanation of what may have occurred. We may start with a group like the tunicates. They represent more or less the "end of the line" for the sessile filter feeders. The next group are the relatively advanced mobile cephalochordates. What lay between? Our "missing link," or transitional animal, can be seen in tunicate larvae, some of which are tadpole like. Paradoxically, in the progression toward a free-swimming chordate, these tiny tadpoles rank higher than their own parents. What if it were possible for these larvae to sexually mature and reproduce tadpoles which would not develop morphologically into sedentary adults? If this could occur, and according to recent studies it appears very possible, then the previous adult stage might fade into obsolescence and the evolutionary progression toward vertebrates would be initiated with the juvenile form. Paedomorphosis, the attainment of reproductive capacity by immature animals which simultaneously abbreviates other developmental traits, is indeed known in several invertebrate groups.

Thus, it appears not unreasonable that the first true vertebrates sprang from the developmentally abbreviated but sexually mature larvae of an ancient pretunicate. In this kind of larva might emerge such important vertebrate characteristics as a larger, anterior gill structure, followed by a muscular tail. The tail in turn is strengthened by the flexible rodlike notochord. Overlying the notochord is the longitudinal nerve cord which extends forward to the anterior sensory organs. Earlier, we noted the presence of these organs in cephalochordates like *Branchiostoma.* It is but a short step beyond the paedomorphic tadpole to advanced fishlike cephalochordates. Possibly, these surpassed modern specialized forms in possessing a rudimentary brain and more highly developed sensory organs. The story has great appeal because of its simplicity. Yet one must be cautious and regard it as a tentative explanation which will probably be modified as evidence continues to accumulate. We have made assumptions based on living animals and lack fossil proofs. We have assumed paedomorphosis in ancient tunicates, yet the living tunicate larva matures quickly into a sedentary sessile animal. It does not feed, nor does it experience gonad development. But then, the explanation favored here does not seek to derive vertebrates from modern animals, but from tunicates or near tunicates that lived several hundred million years ago. We become more firmly convinced when we observe all the right qualifications in the larvae of the living tunicates.

The next great step in this evolutionary journey crosses the boundary between cephalochordates and vertebrates. It came with the development of the vertebral column, and it came without being noticed by the multitude of shellfish.

SUMMARY

Chordates are distinguished from the rest of the animal kingdom by the possession, at least at some stage in their development, of a notochord, a hollow dorsal nervous system, gill slits, a tail, and blood which moves forward in the primary ventral blood vessel and backward in the dorsal. In studies of vertebrate origins, it is necessary to begin with the Phylum Protochordata before proceeding to the more advanced Phylum Chordata. The three classes of protochordates are the Enteropneusta, Pterobranchia, and extinct Graptolithina. Each of these classes has developed one or more traits that either directly or indirectly indicate kinship to the chordate-echinoderm line.

All three subphyla of the Phylum Chordate—namely, the Tunicata, Cephalochordata, and Vertebrata—have a more obviously developed dorsal longitudinal rod. In the Subphylum Vertebrata, the notochord is supplemented or replaced by vertebrae.

In very general terms, the evolutionary progression toward the vertebrates may have begun with sedentary, filter-feeding animals with exposed external cilia located along tentacles. From such a beginning, there may have evolved filter feeders with cilia brought into the body of the animal as gill structures. Gaps in the sides of the pharynx may have opened so that water could escape. In the next stage, free-swimming gilled filter feeders not unlike some modern cephalochordates appeared. With the advent of vertebrae, the last step in this theoretical progression toward true vertebrates was achieved.

SUPPLEMENTARY READINGS AND REFERENCES

Barrington, E. J. W. 1965. *The biology of hemichordata and protochordata.* Edinburgh: Oliver and Boyd.

Berrill, N. J. 1955. *The origin of the vertebrates.* Oxford: Clarendon Press.

Bohlin, B. 1950. The affinities of graptolites. *Bull. Geol. Instit.* 34: 107–113.

Goodrich, E. S. 1958. *Studies on the structure and development of vertebrates.* vol. 1, pp. 1–485. New York: Dover Publications.

Halstead, L. B. 1968. *The pattern of vertebrate evolution.* San Francisco: Freeman.

Kozlowski, Roman. 1966. On the structure and relationships of graptolites. *Jour. of Paleo.* 40 (3): 501–509.

Romer, A. S. 1967. Major steps in vertebrate evolution. *Science* 158 (3809): 1629–1637.

9 Fishes

Shoals of *Cephalaspis,* with their broad arrow-like heads and slender angular bodies, feathered with fins, swept past like clouds of crossbow bolts in an ancient battle.
—Hugh Miller, *The Old Red Sandstone*

WHEN AND WHERE

We are accustomed to using the word "fishes" for all primarily aquatic types of vertebrates. However, fishes actually include at least five distinct classes. There are the jawless fishes, or Agnatha, which are represented today by the unlovely hagfish and lamprey, and in the geologic past by armored forms named ostracoderms. The next two classes, the Acanthodii and Placodermi, had evolved bone-supported, hinged jaws. Although abundant in Silurian and Devonian, they are now totally extinct. The remaining two classes consist of the Chondrichthyes and Osteichthyes. Cartilagenous skeletons are the trademark of the Chondrichthyes, whereas the skeletal material in the Osteichthyes consists largely of bone.

We are unable to say precisely when fish began to populate the earth's water bodies. Fragments of scales and bony armor have been found in Lower Ordovician rocks of Russia, and in the Middle Ordovician Harding Sandstone of Colorado. We can therefore be certain about the existence of fish at least about 480 million years ago. One group of fossils, if indeed they are remains of what we are loosely calling "fishes," may extend the date for the first appearance of vertebrates back into the Late Cambrian. These fossils are collectively termed conodonts (fig. 9.1). Many of them resemble tiny fish teeth, and all are composed of calcium phosphate, the usual hard tissue of vertebrates. A large group of conodonts were constructed of concentric laminations like those found in

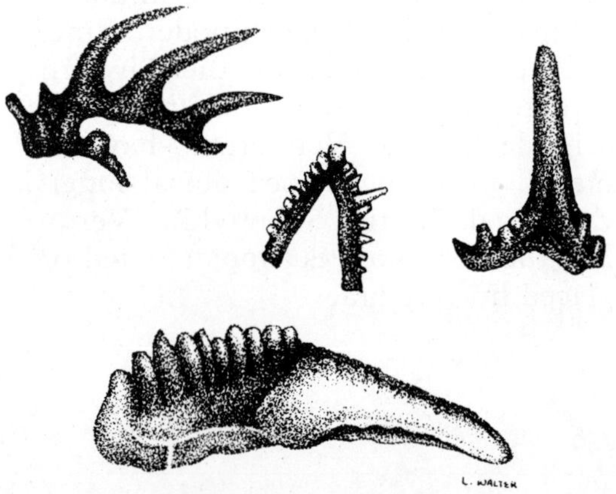

FIG. 9.1 *Conodonts.*

primitive fish scales. Bonelike basal attachments similar to those in fossil fish teeth are commonplace in conodonts. Assemblages appear to represent paired groups and may have been part of the animal's feeding mechanism. As conodonts are found in sediments deposited in a variety of marine environments, they are probably remains of swimming or floating animals. Halstead (1968) has suggested that conodonts "be tentatively interpreted as representing the first, primitive filter-feeding, planktonic vertebrates."

The question of a marine or freshwater birthplace for the fishes is not yet completely resolved. Homer Smith (1963) has argued that fish originated in freshwater bodies and then migrated (in the evolutionary sense) to the sea. According to this concept, once in the sea these fishes would need to eliminate excess water which would enter the body by osmosis —their internal fluids being less salty than the seawater. The glomerulus kidney evolved in order to accomplish this function and is taken as evidence of the transition. It seems unlikely that such an organ would develop in direct response to need, since ordinarily evolution proceeds with adaptations of organs already developed but for a different function. Zoologists also argue that the glomerulus is found in marine animals, as in the hagfish, which probably never had freshwater ancestors. Many believe that the original function of the kidney was that of filtration under hydrostatic pressure and not osmoregulation at all. From the geologic point of view, there is evidence suggesting that the sea was the cradle of fish evolution. The fragmental remains of the oldest vertebrates come from a glauconitic sandstone in Russia. Glauconite, a silicate mineral containing potassium and iron, is universally recognized as forming only in the marine environment. In addition, recent studies of the Harding Sandstone indicate it was primarily a marine, shallow offshore deposit. Thus it appears that, although the question of marine or nonmarine vertebrate origin cannot be laid to rest, the concept of an origin in the sea is favored. After Ordovician time, fishes apparently moved into brackish and freshwater environments, for most of the armored ostracoderms are taken from nonmarine rocks.

OSTRACODERMS

Apart from the previously mentioned fragments of bone and scale from Ordovician rocks of Russia and the United States, the oldest fossils to which formal names can be given come from Silurian strata in England. Two genera, *Jamoytius* and *Thelodus* (figs. 9.2 and 9.3), are found in marine sedimentary rocks. Both genera are extinct representatives of the class Agnatha, the jawless fish. Living members of this group include the lamprey and hagfish. In reference to the armorlike covering of bony scales and plates, fossil agnathans are loosely termed ostracoderms ("shell skins"). Many ostracoderm characteristics might be considered predictable for very primitive vertebrates. All had eyes, ears, and a single nostril on the top of the head. Most possessed a strong propulsive tail which was commonly upturned (heterocercal) and some, more rarely, downturned (hypocercal). The head region was often expanded and contained a large gill chamber as well as the usual sense organs and blood vessels. In most groups, the jawless, usually toothless, mouth was located ventrally. A flattened shape completes the image of a bottom-dwelling animal that fed on food particles strained from the soft sediment, or sucked in a current of water from which food was extracted. In some forms, there were a pair of laterally projecting anterior appendages which lacked recognizable internal bony supports. Many ostracoderms, as exempliefied by *Cephalaspis* and *Hemicyclaspis* (fig. 9.4), possessed prominent head shields and scales of bone. The function of the armor, which has become the hallmark of the ostracoderms, may have been for protection against predators, possibly the larger cephalopods of the Early Paleozoic. A less-favored possibility is that dermal armor arose as a covering which functioned in impeding osmotic water transfer in primarily freshwater ostracoderms that "went to sea." If fish did not originate in freshwater bodies, then this concept would be invalidated. If they first developed in streams and lakes, then the development of armor would appear to indicate these fish were preparing themselves for a future different way of life. There was no piscatorial crystal ball, and this sort of predisposition is not likely to occur in evolution except in the sense of modifying structures for new functions. Another theory for the early development of dermal armor explains its origin as storage banks for seasonally available phosphorous. According to Pautard (1961), phosphates could be stored by the ostracoderms during times

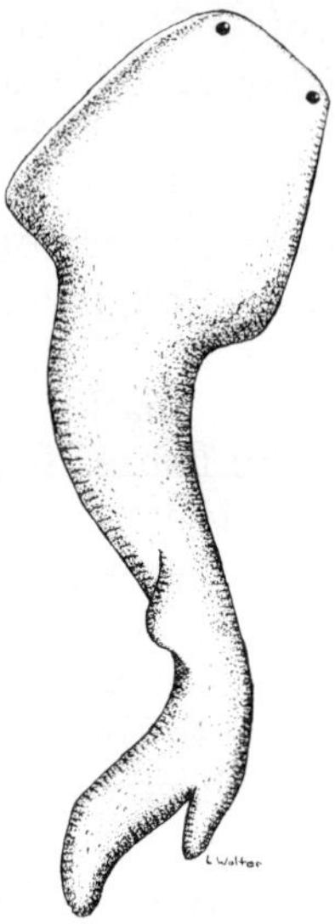

FIG. 9.3 Thelodus.

of seasonal upwelling and utilized during times of seasonal deficiency. This handy store might have been important to organisms like the early vertebrates, that required a consistent source of phosphorous to maintain a suitable level of mus-

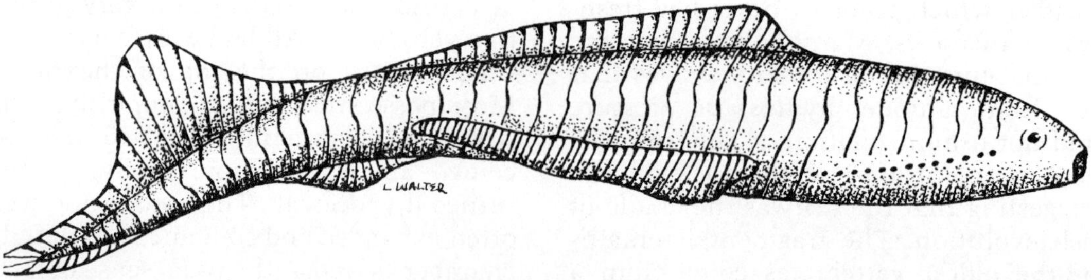

FIG. 9.2 Jamoytius.

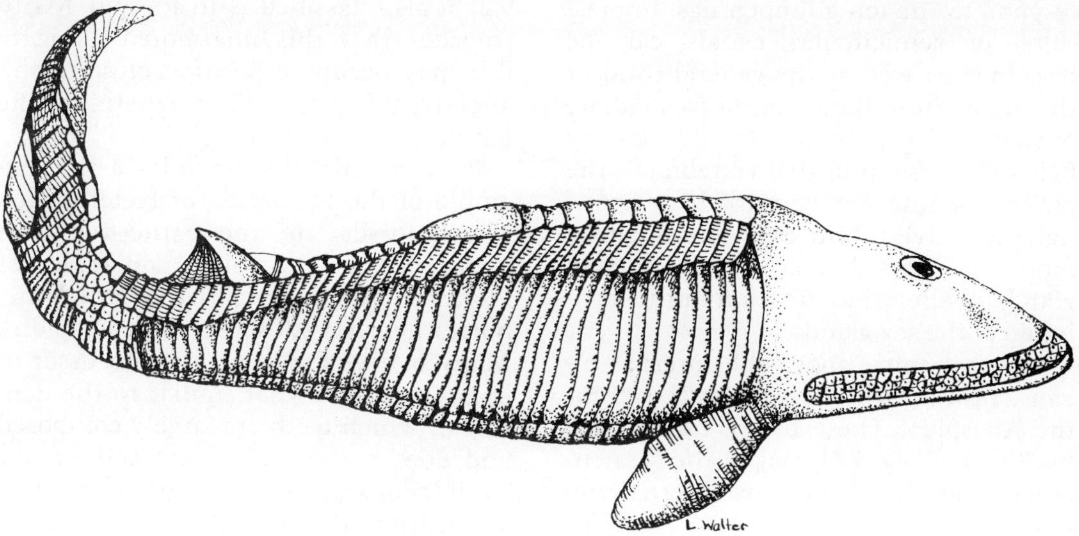

FIG. 9.4 Hemicyclaspis.

cular actvity. The fact that the bony plates provided protection also, may have been an important evolutionary bonus.

The ostracoderms lived from the Ordovician to the latter part of the Devonian. They were a diverse group and, because many orders of ostracoderms were nearly contemporaneous, it is difficult to trace their phylogenetic lines. Four orders, the Cephalaspida, Anaspida, Pteraspida, and Coelolepida, are relatively well known.

Cephalaspis (fig. 9.5) is perhaps the most famous of all ostracoderms. It is readily recognized by its large, flattened, semicircular shield on which can be found two eye openings, a small pineal opening, and a single nostril. Depressed areas along the lateral margins of the shield may have had a sensory function. Propulsion in swimming was accomplished by a heterocercal tail suitable for driving the animal slightly downward in the water as befits a bottom dweller. The flat-bellied body was completely covered with rectangular bony plates. The fish rarely exceeded thirty centimeters in length.

The interior of the cephalic shield of cephalaspids contains impressions of soft parts. Careful study has delineated a regular segmental arrangement of nerves,

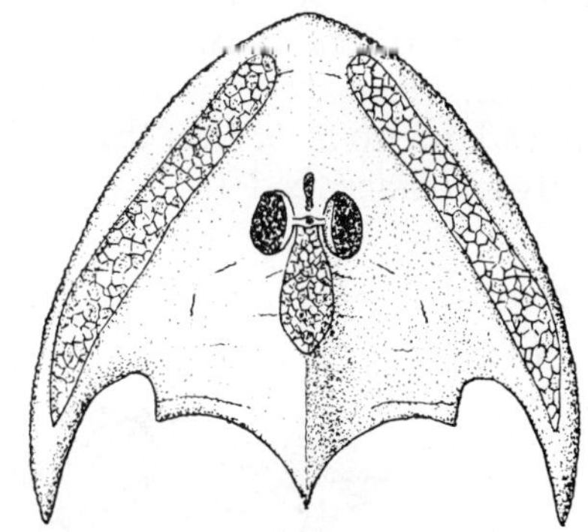

FIG. 9.5 Cephalaspis. *(Modified from J. A. Moy-Thomas, 1971,* Palaeozoic Fishes, *2d ed., Chapman & Hill, Ltd., London, p. 14.)*

blood vessels, and gill pouches that correspond to the ten gill openings. Impressions of semicircular canals can be discerned as well as the general form of the brain. From the brain, nerves radiate outward to supply the lateral sensory fields. In these primitive vertebrates, the pineal opening was very likely a sort of light sensitive "third eye," although its more modern derivative, the pineal gland, is believed to function in the regulation of the sex glands.

Less is known about the internal organs of the second order of ostracoderms, the Anaspida. These fishes, exemplified by *Birkenia* (fig. 9.6), lacked the semicircular head shield with its internal im-

FIG. 9.6 Birkenia. *(After A. S. Romer, 1966,* Vertebrate Paleontology, *3d ed. University of Chicago Press, p. 19.)*

pressions of soft parts. *Birkenia* lived differently than *Cephalaspis.* The downturned tail, deep body, and terminal mouth lead one to infer that this fish was primarily a surface feeder.

A very unusual fossil, *Jamoytius* (fig. 9.2), is also classified as an anaspid. Many consider that this unarmored elongate fish may occupy a position close to the ancestry of those living agnathids, the lamprey and hagfish.

Pteraspis (fig. 9.7) provides a fine example of the Pteraspida or Heterostraci. These fossils are the earliest known ostracoderms. The shield and scales of *Pteraspis* are not composed of true bone, but rather of a substance called aspidin. Many believe aspidin is the precursor of bone. It is somewhat similar to the dentine of which teeth are largely composed and does not contain bone cell spaces. In slides of aspidine tissue, one can detect a transition from acellular calcified tissue to hard tissue that truly resembles cellular bone. The concept that the earliest hard tissue among vertebrates was aspidine and that bone evolved from it, is also supported by the fossil record. The previously mentioned, fragmental fish remains of the Ordovician are of aspidine composition and presumably came from pteraspids. Thus, it seems certain that vertebrates acquired bone at a very early stage in their history.

The final order of ostracoderms, the Coelolepida, did not lend themselves readily to fossilization. The body of these fishes was covered with minute scales rather than shields. This, of course,

FIG. 9.7 Pteraspis.

greatly reduced the probabilities for preservation. *Thelodus* (fig. 9.3) is a suitable representative.

THE INNOVATED HINGE

Ostracoderms did not survive beyond the Devonian. For the most part, they had been small, rather frail creatures restricted by their mud-straining or filter-feeding mode of life. They were surpassed and gradually replaced by the jawed fishes. We presume that the forebearers of the first jawed fishes were ostracoderms, but the precise ancestor has not been determined. The ostracoderm fossils in museum collections are in relatively advanced states of specialization. The remains of the fishes that link the ostracoderms to the jawed fishes probably lies somewhere in strata of the Cambrian or Ordovician.

The appearance of bone supported true jaws was without doubt one of the most outstanding events in the history of vertebrate evolution. The ability to bite, grasp, and manipulate led to new, more varied, and more active ways of life. Fish became browsers on aquatic plants and predators on animals far too large for the agnathans to consume. The fish became the unchallenged rulers of the seas.

The story of the origin of the jaw is a truly remarkable account of the adaptation of a structure so that it can perform a role far different than its original function. The original structures were gill supports. The evidence of embryology, paleontology, and homology combine to demonstrate that certain of these supports evolved into the vertebrate jaw.

Earlier we noted the presence of gills in prevertebrate chordates. It is likely that many of these animals also possessed cartilaginous or bony supports between gill openings. Such supports, for example, are present in the living *Branchiostoma*. The gill-supporting components in primitive vertebrates were composed of a sequence of rods arranged so as to form V-shaped structures. These were turned on their sides with the point of the V directed toward the rear (fig. 9.8). A series of these sidewise V's were arranged on either side of the throat region. As the mouth enlarged and extended backward, one or two of the anterior pairs of V's were eliminated. The next pair of V-shaped gill supports are of special interest. These enlarged and the lower arms of the V became freely moving in hinge-like fashion. Thus a pair of gill supports were transformed into jaws. At this early stage of development, the upper jaw (palatoquadrate) articulated with the overlying braincase. A full gill slit lay behind the jaw. In subsequent evolution, the next most posterior pair of gill supports (hyoid arch) was modified so that the up-

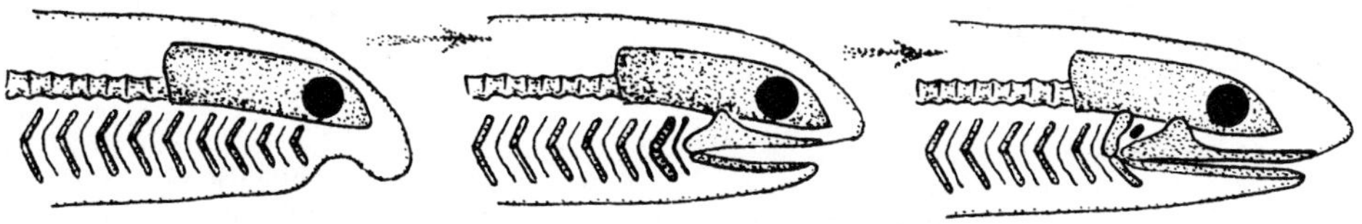

FIG. 9.8 *Very generalized diagrams depicting the origin of the jaws from gill arch supports. (Modified after J. M. Weller, 1966,* The Course of Evolution, *McGraw-Hill, p. 390.)*

per member (hyomandibular) formed a support between the jaw and the braincase. The pair of gill slits that once existed between the jaws and what had become the hyomandibular were squeezed into a smaller space and ultimately developed into the pair of circular spiracle openings. The spiracles were destined to become a part of the auditory apparatus in later animals, and portions of the gill arch behind the jaw were to function in the support of a moveable tongue.

The origin of the vertebrate jaw was first perceived over eighty years ago. Since then, zoologists have noted that the jaw elements in sharks were in line with, and similar to, the gill supports. In the jargon of biologists, the jaws, the hyoid arch, and the gill arches formed a homologous series. Even the spiracle was clearly a serial homologue of the gill slits. Of even greater significance is the fact that the nerves and major blood vessels of the jaw have an arrangement obviously similar to their arrangement in the gill apparatus. For example, nerve communication to each of the gill pouches consists of a single nerve which divides into two strands which straddle the gill pouch, and one which extends forward over the alimentary canal. The posterior of the pair of nerves that straddle the gill pouch serves to transmit muscle-contracting impulses. In direct comparison, this posterior nerve in the jaw transmits impulses to muscles which move the jaw. The other, or anterior nerve, provides for taste and smell.

NEW DESIGNS

For a time the ancient jawed fishes lived side by side with the ostracoderms. By Devonian, jawed fish dominated, and the agnathids, perhaps because of unfavorable competition and predation, dwindled to final extinction at the end of the period. It was the heyday of fishes, and they radiated into a variety of marine habitats. Some retained or regained the flattened body and heavy armor reminiscent of the bottom-dwelling ostracoderms. Others became streamlined and either lost or reduced the covering of cranial armor. Perhaps the more active habits and mobility of the new fishes made heavy or excessive armor unnecessary. Continuous lateral folds to assist in swimming appeared in some fish designs. Others developed rows of paired spines which may have supported flaps of dermal tissue. In some of the more sluggish looking forms, powerful arthropod-like paired appendages are developed just behind the head. It was a time of change and experimentation.

The earliest fishes with moveable lower jaws were the acanthodians. They first appear in nonmarine rocks of the Late Silurian, become most abundant during the Devonian, and then decline toward Permian extinction. *Acanthodes* (fig. 9.9) is a representative of the earliest

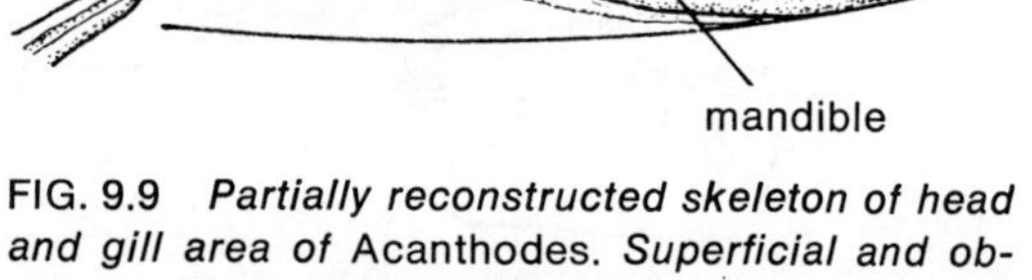

FIG. 9.9 *Partially reconstructed skeleton of head and gill area of* Acanthodes. *Superficial and obscuring structures are not shown. (Modified from A. S. Romer, 1966,* Vertebrate Paleontology, *3d ed. University of Chicago Press, p. 35.)*

acanthodians. A lower jaw, or mandible, articulates with the primitive upper jaw (palatoquadrate). Behind the jaw, the upper bone of the hyoid arch (the hyomandibular) was somewhat enlarged and sturdily propped against the palatoquadrate. A full-sized gill slit still remained between the jaws and hyoid arch. Evolution had not yet progressed sufficiently to produce the spiracle. Thus, *Acanthodes* nicely shows a stage in the sequence of steps that produced jaws from gill supports.

Climatius (fig. 9.10) is a somewhat more advanced acanthodian. One is impressed with its tapered, fishlike body and appearance of mobility. Only a few inches long, *Climatius* had large eyes surrounded by a ring of protective bony plates. The forward position of the eyes and lack of a long snout imply that sight was far more important than the sense of smell in this fish. There were a series of spines along the lower sides of the fish which may have supported folds of skin. A heterocercal tail and body covered with tiny rhombic- or diamond-shaped scales characterized these acanthodians.

Another group of archaic jawed fish collectively known as Placoderms is known from Late Silurian through Permian strata. Placoderms offer many examples of adaptational variations. One such example consists of the joint-necked arthrodires. These fishes possessed a pair of moveable joints which separated the head shield from the carapace covering the forward thoracic region. The joints permitted the simultaneous raising of the skull and dropping of the lower jaw so that the mouth could open widely to engulf large prey. The predatory habits of the arthrodires is further evidenced by the peculiar sharp plates in the jaws which served to snare and cut the prey. *Coccosteus* (fig. 9.11) is a representative arthrodire whose length rarely exceeded a meter. At the opposite extreme in size, was the huge and formidable *Dinichthyes* (fig. 9.12) which grew to lengths exceeding nine meters. All of the arthrodires lacked complete centra in the vertebral column. Arthrodires had paired spines

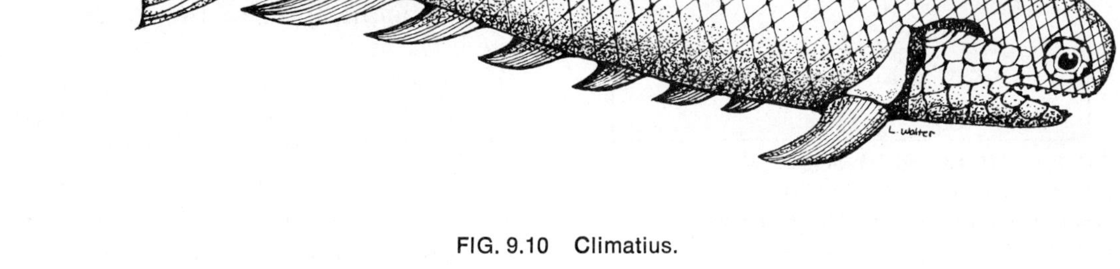

FIG. 9.10 Climatius.

behind the cephalic shield which may have supported pectoral fins. In addition, there were a pair of pelvic fins supported by radial rods and a heterocercal tail.

One might facetiously suspect the group of placoderms known as antiarchs to have yielded to evolutionary nostalgia.

FIG. 9.11 Coccosteus. *(Modified after E. H. Colbert, 1969,* Evolution of the Vertebrates, *2d ed., Wiley, p. 37.)*

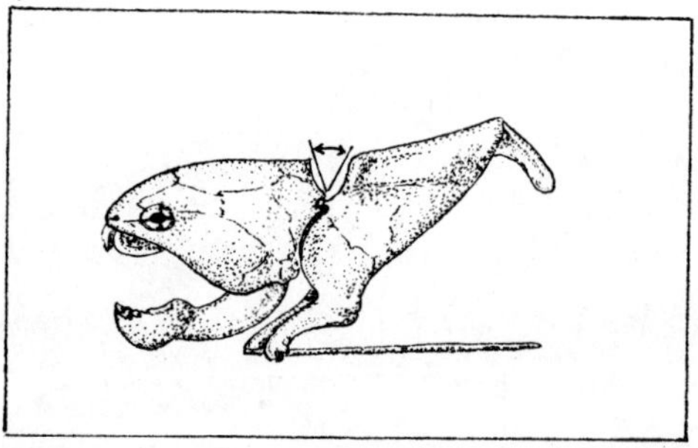

FIG. 9.12 *Head and thoracic shield of the giant Devonian placoderm* Dinichthys. *Skull is over five feet in height. Notice the hinge between the head and thoracic shield.*

At first glance they appear to have made a return to the heavily carapaced form of the mud-grubbing ostracoderms. They are, however, only the product of adaptation to a similar mode of life. The tendency of a group of animals to develop general similarities to distantly related groups as a result of exploitation of similar habitats is termed convergent evolution. It is a common phenomenon in the overall history of life. Actually, the antiarchs were highly specialized placoderms, probably derived from an ancestral arthrodire. They were small fish of the Middle and Late Devonian. *Bothriolepsis* (fig. 9.13) is a well-known fossil antiarch. The head and forward parts of the body were covered with imbricating plates arranged in a regular and consistent pattern. As in many placoderms, the plates contained bone cells and were constructed of three layers. The basal layer consisted of laminated bone, the middle layer of trabecular bone, and an upper layer of laminar bone often supporting tubercles or grooves of the lateral sensory system. The eyes of antiarchs were located on either side of a pineal opening. Two anterior notches have been interpreted as nostrils. Some fossils of antiarchs even exhibit the phantom traces of internal organs believed to represent air sacs or lungs. This leads one to a favorable view of the hypothesis that lungs evolved early in fish evolution, and that the hydrostatic swim bladder found in more recent fishes represents a modification of the lungs for a different function.

If the antiarchs remind one of "evolutionary throwbacks," then the placoderms called rhenanids suggest evolutionary prefigurements of some modern members of the shark family. *Gemuendina* (fig. 9.14), for example, appears to

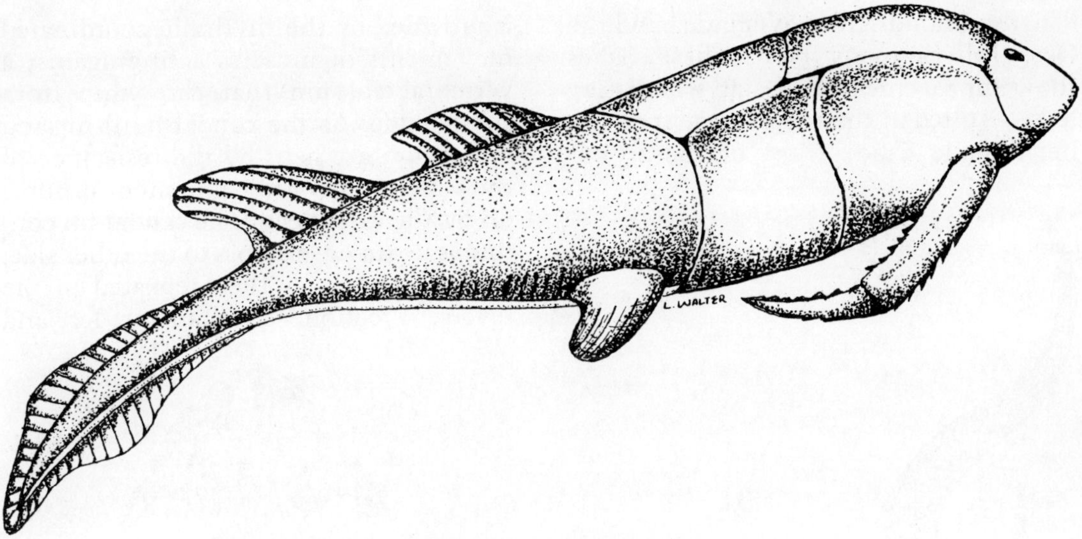

FIG. 9.13 Bothriolepis.

anticipate the form of the rays and monk fishes. However, on closer examination it is apparent that the sharklike characteristics are only superficial. Rhenanids possessed bony plates on the anterior parts of the body, and bone is a tissue absent in true sharks. In many details, the Early Devonian rhenanids were more advanced than the relatively primitive earliest sharks of the Late Devonian. An ancestor should be more primitive than his supposed descendents. The rhenanids appear to have been operculate fishes, and sharks do not have the opercular covering over the gill apparatus. Even the jaw suspension of the rhenanids was unlike that found in sharks. It appears to be another case of convergency. Perhaps like the modern rays, *Gemuendina* used their expanded pectoral flaps to flip a camouflaging film of sediment onto their backs as they lay watchfully in wait of prey.

Petalichthyids, represented by the odd *Lunaspis* (fig. 9.15), were rather specialized, small fishes with head and trunk armor. Paired pectoral fin-spines, two dorsal fins, and diphycercal tails characterized the petalichthyids.

Incertae sedis is the designation applied (with a small measure of frustration) to the problematic palaeospondylids. The term means "position uncertain," and this is indeed appropriate, for these puzzling fossils have variously been considered agnathans, larval hagfish, larval lungfish, larval amphibians, or archaic jawed fish. Rarely exceeding fifty millimeters in length, *Palaeospondylus* (fig. 9.16) possessed a well-calcified skull, rather advanced vertebral column with distinct centra, and a prominent heterocercal tail. There were also traces of pelvic fins. They are placed here among the jawed fishes because they may have had jaws, or at least structures resembling the palatoquadrate and mandible.

TRENDS IN FIN AND SCALE

The culmination of fish evolution is seen in the two great classes of higher fish, the Chondrichthyes (sharks) and the Osteichthyes (bony fish). These fishes show improvements in design which may have resulted in their replacement of earlier archaic fishes. Most of these "improvements" related to greater efficiency and speed in swimming. An arrow-shaped head cleaves the resistant water. The head is followed by a streamlined body that offers a minimum amount of resistance to forward motion. Propulsive force is provided by the rhythmic coordinated movements of muscles acting against a vertebral column that can whip from side to side. As the caudal (tail) fin and tail region are bent by the resistance of the water, a forward component of force drives the fish ahead. The caudal fin continues its traverse across to the other side, and the motion is then repeated in the reverse direction. Simultaneous left and right movement of the head region is negligible because of the resistance which the forward part of the body offers to the flow of water lateral to its surface, and because of the relatively high moment of inertia produced at the forward end of a rapidly moving fish. Fishes of ordinary size like trout have been observed to swim at maximum speeds of six miles per hour. The principal muscles providing this movement and their relationship to the internal skeleton are indicated in figure 9.17. Note that in the region anterior to anus, the muscles thin, and the haemal

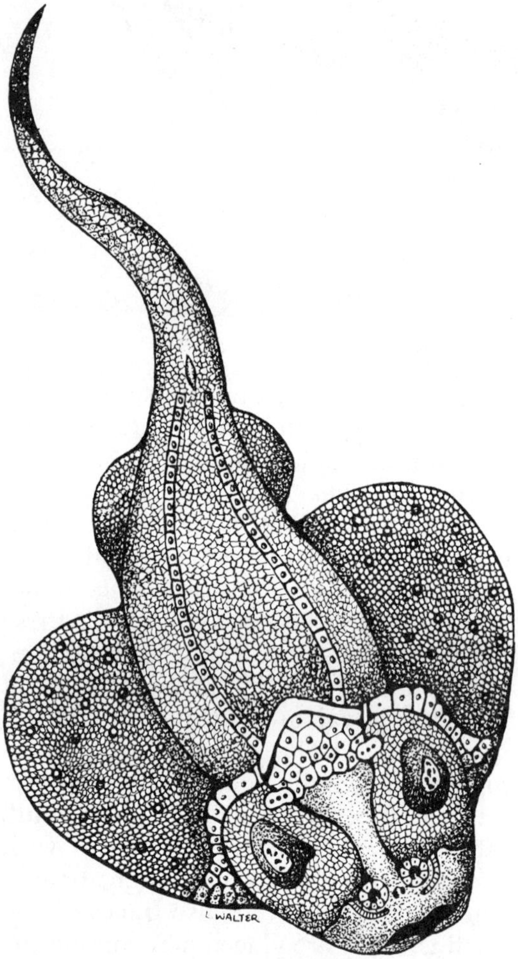

FIG. 9.14 Gemuendina.

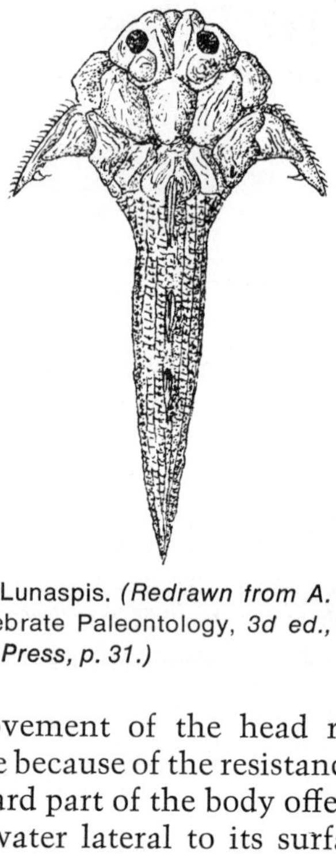

FIG. 9.15 Lunaspis. *(Redrawn from A. S. Romer, 1966,* Vertebrate Paleontology, *3d ed., University of Chicago Press, p. 31.)*

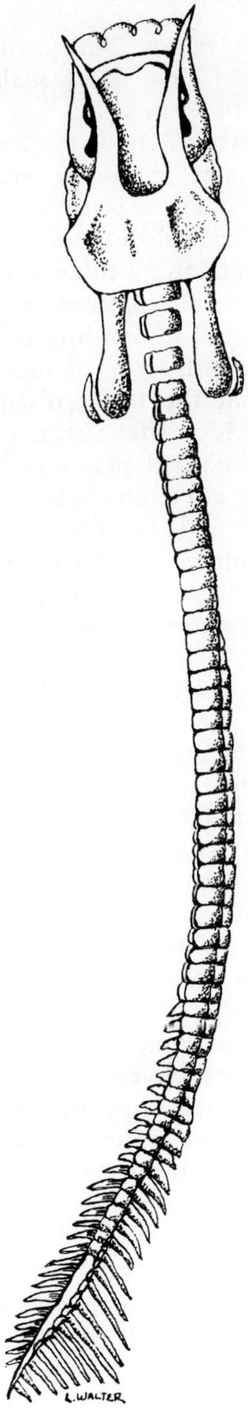

FIG. 9.16 Palaeospondylus.

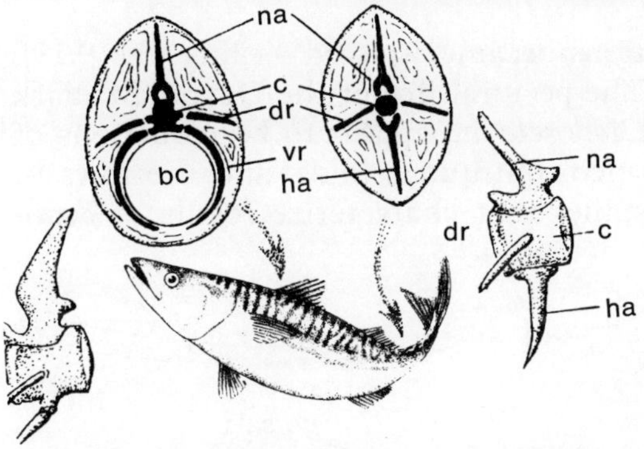

FIG. 9.17 *Schematic section through a fish. Abbreviations:* bc, *body cavity;* c. *centrum;* dr, *dorsal ribs;* ha, *haemal arch and spine;* na, *neural arch and spine;* vr, *ventral ribs.*

arches lose their spines and open to admit the visceral organs. In addition to the caudal fin of fishes paired fins provide turning-braking movements. Like the ailerons and elevators of aircraft and submarines, they prevent roll, pitch, and yaw (fig. 9.18). The keellike dorsal fins

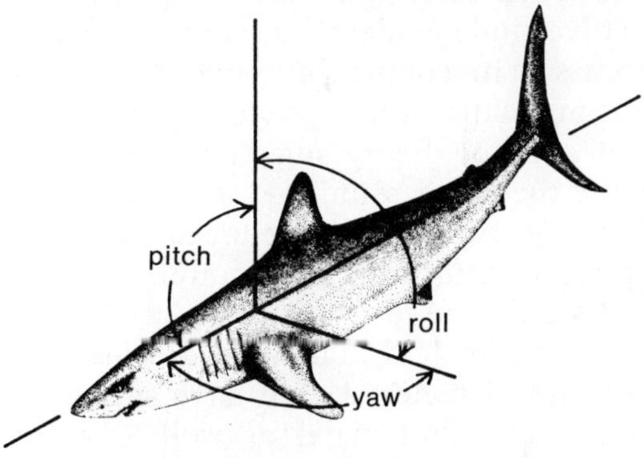

FIG. 9.18 *Pitch, roll, and yaw.*

on the dorsal midline of the fish assist in these functions and help to prevent side slip.

Fins began their evolution among the archaic jawed fishes. A popular hypothesis states that fins were derived from thin continuous folds, and that as portions of the fold disappeared, the persisting portions became isolated as fins (fig. 9.19). The pectoral fins of the Devonian shark *Cladoselache* appear to be reductions of once continuous folds, and the rows of spines that characterize fossils of acanthodians may have supported lateral folds in life. Another possibility is that only the pectoral fins were developed from a shortened fold, and that the pelvic fins evolved independently as clasping organs to assist in copulation. Indeed, such is the main function of pelvic fins in sharks. Yet another theory suggests that spines were the more primitive, and that later flaps of skin were produced between spines transforming the combined structures into true fins.

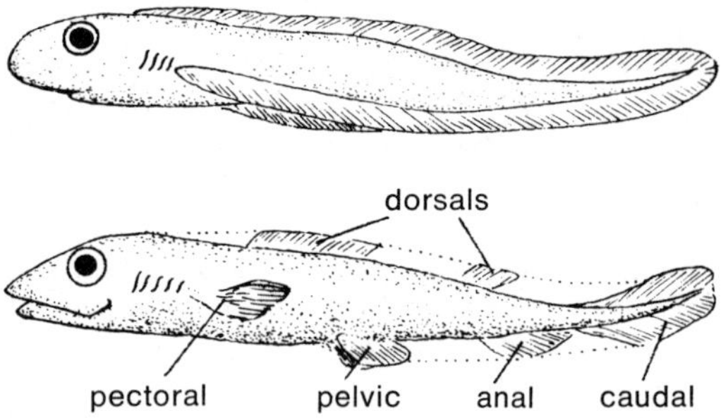

FIG. 9.19 *Theoretical derivation of fins from lateral fin folds such as those on* Branchiostoma. *(Modified after Storer, Usinger, and Nybakken, 1968,* Elements of Zoology, McGraw-Hill, *p. 384.)*

Scales, no less than fins, are obvious features of fishes. As we have seen, not only scales, but plates as well, covered parts of the bodies of ancestral fishes. One is led naturally to speculate on the evolutionary relationships of these two coverings. One line of speculation has led to the concept that plates are a more advanced feature resulting from the growth and fusing together of many individual scales. Quite the opposite is also a possibility, and scales may represent small elements that regressively replaced earlier bony plates. The former concept which would have small, scalelike units coalescing to form plates has an appealing simplicity. Careful studies of ostracoderms have shown a sequence of developmental stages from isolated tubercles (prescales) in the skin, to peripheral growth of these tubercles until they coalesced into a pattern of hexagonal units (tesserae) resembling terrazzo, and ultimately confluence of individual tesserae so that the terrazzo aspect disappears. The idea that scales or scalelike tubercles preceded plates is further enhanced by our knowledge that oldest vertebrate remains do not consist of well-preserved plated animals, but only of great numbers of small tuberclelike bone fragments scattered throughout the host rock.

Scales of many ostracoderms had an inner portion that was composed of soft laminar bone covered by a harder material lacking in blood vessels that somewhat resembled dentine. A still harder "enameloid" substance formed nodes or ridges on the exposed outer surface. With this hard outer layer, the overall structure resembled that of the placoid scale of sharks. A reasonable explanation appears to be that dermal denticles of sharks were derived from somewhat similar structures developed by agnathan ancestors. Following the evolutionary story one step further, it is recognized that sharks' teeth are themselves specialized scales. As one examines the scales across the "lip" area of sharks, one can see that placoid teeth form a continuous homologous series with the scales. More generally, teeth in all vertebrates appear to be ultimately derived from scales or scalelike dermal elements of Paleozoic fishes.

The scales of bony fishes show many interesting variations. Two types appear characteristic of the more ancient osteichthyids. One of these, termed cosmoid, was prevalent among the fish ancestors of amphibians (crossopterygians). The bone in these cosmoid scales consisted of two layers, a dentinelike cosmine layer with branching tubules underlain by more normal bone called isopedine. Between these layers was an area much like a pulp cavity. The surface of the scale was encased in a hard layer resembling enamel and the overall shape was rhombic. The second type of primitive fish scale is called ganoid and persists today in the sturgeon. The general shape and structure of the ganoid scale is similar to the cosmoid except that the enamellike surface is deposited in thick layers called ganoine, and the scale grows by additions to its surface as well as its edges.

In most modern bony fishes, both of these earlier types of scales have been replaced by a type of scale called cycloid (fig. 9.20). It consists of roughly circular thin discs of bone which in life were overlapping posteriorly and covered by a thin layer of skin. Pigmented cells in the skin layer provide color markings in living fish. Annular rings representing growth lines are characteristic of cycloid scales. In some fish scales, the posterior edge develops rows of tiny points producing a modified cycloid scale termed ctenoid.

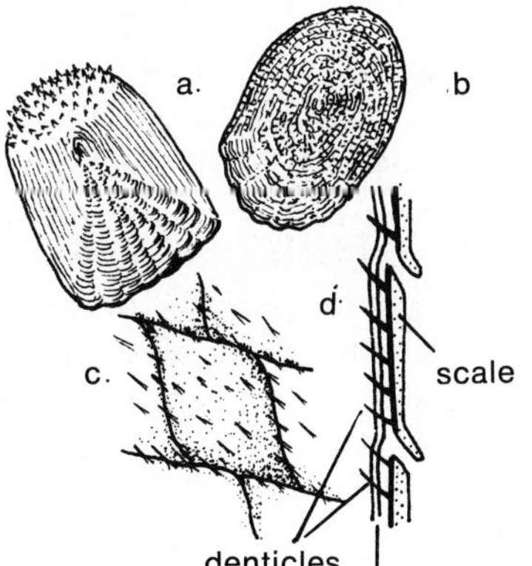

FIG. 9.20 *Scales of bony fishes (enlarged);* a, *ctenoid scale with tiny teeth;* b, *cycloid scale;* c, *ganoid scale in surface view;* d, *cross section through a ganoid scale.*

THE CARTILAGINOUS CONTENDERS

It was during the Devonian Period, aptly called the "ages of fishes," that the cartilaginous fishes, or chondrichthyoids, made their debut. Members of the Orders Cladoselachii, Pleuracanthii, and Bradyodonti were common during the Late Paleozoic as evidenced by the frequency with which their teeth are found in strata. Living chondrichthyoids include the sharks (Selachii), skates and rays (Batoidea), and the rat fishes (Chimerae).

As the name implies, chondrichthyoids have cartilaginous endoskeletons. It is uncertain whether or not they evolved from cartilaginous ancestors, or whether they branched from ancestral bony fishes and became secondarily cartilaginous. Whatever its origin, the cartilaginous condition provides for poor preservation. Teeth, placoid denticles, and spines comprise the bulk of fossil material. The denticles cover the exterior of the body and impart a sandpaper "feel" to the surface of sharks. The denticles are formed in the skin (fig. 9.21). Bone-forming cells in the skin's dermal layer secrete dentine around a pulp cavity while the overlying epidermis covers the dentine with enamel. The apex of the denticle is then pushed through to the surface. As we have previously noted, denticles form a continuous homologous series with the teeth which are merely enlarged denticles. The mouth in living sharks is ventral. The teeth are not fused

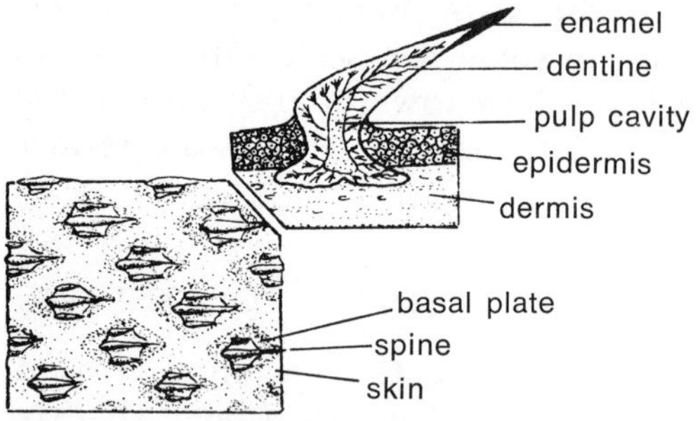

FIG. 9.21 *Surface view of placoid scales and median section through a scale. (Modified from Storer, Usinger, and Nybakken, 1968,* Elements of Zoology, *McGraw-Hill, p. 384.)*

to the jaws, but rather grow from the skin covering the jaws. Teeth in carnivorous forms are used for holding and tearing, but not for chewing. They are formed in rows, and lost or damaged teeth are replaced in series by new teeth rotated into position. Bottom-dwelling chondrichthyoids generally feed upon molluscs. As an adaptation for crushing shells, these fish possessed pavementlike teeth with flat surfaces.

Internal fertilization of eggs is usual in sharks. To aid the sperm in entering the female's reproductive tract, male sharks of most species bear clasping structures in the pelvic region. Except for one order, gills are not covered by opercula, so that the gill slits are visible on either side of the anterior portion of the body. There are no air bladders and no lungs. Members of the shark lineages characteristically have the spiracle—the reduced gill slit discussed earlier in connection with the origin of the jaw. The sense of smell is highly developed in sharks and appears to be relatively more important in the search for prey than sight.

Cladoselache is a Devonian shark that exemplifies the characteristics of the early chrondrichthyoids. Fossils come chiefly from exposures of the Cleveland Shale along the southern shore of Lake Erie. The formation contains elliptical nodules some of which contain remains of teeth, calcified skeletal cartilages, and more rarely skin and traces of body outline and internal organs. Because of this exceptional preservation, the appearance of *Cladoselache* is well known (fig. 9.22).

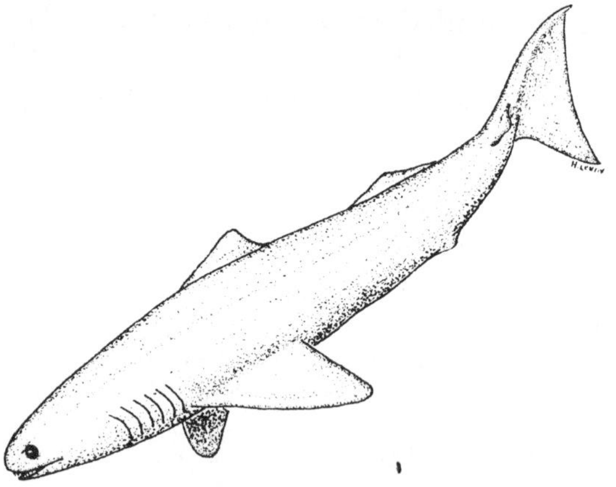

FIG. 9.22 Cladoselache.

The fish had a torpedo-shaped body which was covered by placoid scales and lacked clasping structures. The caudal fin was heterocercal. Paired fins were developed that somewhat resemble the fin folds of living shark embryos. The suspension of the upper jaw was relatively primitive in that it was partially supported by the hyomandibular and was attached also directly to the braincase by two short processes. This type of jaw suspension is termed amphistylic. In modern sharks the suspension is hyostylic and support for the upper jaw is accomplished by the hyomandibular alone. Within the jaws were teeth, each of which consisted of one large central cusp and lateral smaller cusps. These teeth are common fossils and attest to the abundance of cladoselachians.

Sharks have been predominantly marine animals throughout their history, but one ancient group did manage to penetrate the freshwater environment. These nonmarine sharks are represented by *Xenacanthus* (fig. 9.23). *Xenacanthus*

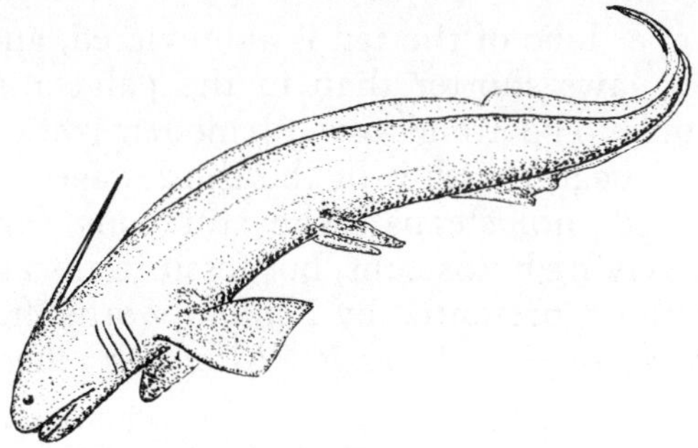

FIG. 9.23 Xenacanthus.

had a diphycercal tail, a long keellike dorsal fin, a prominent spine on the back of the head, and the males possessed clasping organs. The teeth were distinctive in having two prominent lateral cusps on either side of a tiny central denticle. Like the cladoselachians, the jaws were not only braced by the hyomandibular but also attached to the forward area of the braincase.

Another Paleozoic group, the bradyodont fishes, are known primarily from their teeth, which were broadly rounded pavement teeth, probably used for crushing shellfish. Like the modern rays, bradyodonts had the flattened bodies and winglike pectoral fins characteristic of fish adapted for life on the seafloor.

Modern sharks arose from cladoselachian ancestors during the transition from Devonian to Mississippian, but remained fairly primitive until Jurassic when the more familiar modern predatory sharks and the mollusc-eating rays and skates became abundant. Many interesting adaptations are found among living species. Some have developed electric organs as defensive and guidance systems. The tail of the rays, no longer required in their life on the sea bottom, evolved into a wound-inflicting spine. Adaptations for snaring animals on sharp teeth or for crushing shells are common, but whale sharks and basking sharks have returned to filter feeding and consume large quantities of marine plankton. The bizarre sawfish shark uses its long, toothlined nose to slash about in schools of freshwater fish. The victims are impaled on the daggerlike teeth and then consumed. The Chimerae, or rat fishes, are also strange looking but comparatively rare fishes that inhabit the abyssal depths. Their stout, shell-crushing dental plates effectively break the exoskeletons of snails and other molluscs. Of all the living members of the cartilaginous fraternity however, the predaceous, powerful, and unpredictable "man-eater" *Charcharodon* is recalled most often by wary man.

THE BONY HOST

The Osteichthyes, or bony fish, are the most numerous, successful, and varied of all the aquatic vertebrates. They have adapted themselves to waters of all kinds; fresh or saline, light or dark, cold or warm. There are even some that are not entirely restricted to water. The "mudskipper" and "walking catfish of Florida" can move about on land. Certain lungfish encase themselves in cocoons and live for months without water. Some fishes like the "flying fish," are even capable of limited gliding flight. Organs that produce light, or venom, or electric currents are also present in some members of the highly varied bony fishes.

In addition to the at least partially ossified skeleton, Osteichthyes have gills covered by an operculum and usually possess swim bladders. Unlike their chondrichthyoid cousins, the males do not possess claspers, nor do they have placoid scales. Osteichthyes are divided into two lesser categories, the numerically important ray-fins (Actinopterygii) and the tetrapod ancestors, the lobe-fins (Choanichthyes).

Ray-fin fishes, as is implied by the name, possess thin fins which lack a muscular base and are supported by radiating bony rays. Unlike the Choanichthyes, these fish have blind pits for nostrils and not paired nasal passages that open into the pharynx. The ray-fins began their evolution in the freshwater bodies of the Devonian and expanded into the marine environment to become the dominant fishes of the modern world. In the course of their evolution, four generalized stages can be recognized. The earliest stage is represented by a group of ancestral bony fishes termed "palaeoniscoid chondrosteans." These fishes had rather incompletely ossified skeletons, as befits the ancestor of bony fish. They are well represented by the genus *Cheirolepis* (fig. 9.24). These fishes had glossy rhombic ganoid scales, long jaws, antique heterocercal tails, and retained the spiracular slit. They gave rise to the "advanced chondrosteans" during the Triassic. *Redfieldia* is representative of this second stage. The scales show a reduction in the middle layer of dentinelike material, the

FIG. 9.24 Cheirolepis.

upper lobe of the tail is abbreviated, and the jaws shorter than in the palaeoniscoids. The trend toward modernization had begun. Fishes of the third stage are called holosteans. The sturgeons are surviving holosteans, but fossil forms are better represented by *Pholidophorus* (fig. 9.25) of the Jurassic. Although these

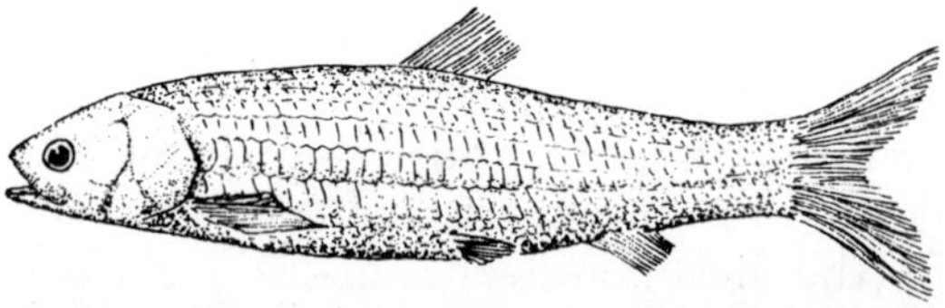

FIG. 9.25 Pholidophorus. *(Modified from E. H. Colbert, 1969,* Evolution of the Vertebrates, *2d ed., Wiley, p. 55.)*

fishes retain the primitive rhombic shape of scales, the upper lobe of the tail is even more abbreviated than in the advanced chondrosteans. The lungs are transformed into an air bladder, and the spiracle is lost. The teleosts comprise the fourth and most advanced stage of osteichthyoid history. These modern fish have completely ossified internal skeletons, thin and rounded scales of the cycloid design, and caudal fin of the advanced homocercal type. In many species, the pelvic fins have migrated toward the anterior. Perch, bass, carp, snappers, swordfish, minnows, herring, and a host of other living teleosts comprise this culminating host of actinopterygians.

THE HARDY AIR GULPERS

The second great category of bony fishes, termed Choanichthyes, are of tremendous evolutionary significance. Their name (Greek *choana,* internal nostril and *ichthyos,* fish) refers to a pair of openings in the roof of the mouth which communicate with the external nostrils. In a way, choanichthyoids could, when necessary, "breath through their noses." Although such fishes are uncommon today, in Devonian times they were very numerous. More important to the history of life, they formed the bridge between aquatic and terrestrial vertebrates. These were the fish that came ashore. Loren Eiseley (1959) nicknamed one of these prophetic Devonian fish "the snout" and offered this imaginative description of its travail.

The pond was a place of reek and corruption, of fetid smells and of oxygen starved fish breathing through labored gills. At times the slowly contracting circle of water left little windrows of minnows who skittered desperately to escape the sun, but who died, nevertheless, in the fat warm mud. It was a place of low life. In it the human brain began. . . . On the oily surface of the pond, from time to time, a snout thrust upward, took in air with a queer grunting inspiration, and swirled back to the bottom. The pond was doomed, the water was foul, and the oxygen almost gone, but the creature would not die. It could breathe air through a little accessory lung, and it could walk. In all that weird and lifeless landscape, it was the only thing that could. It walked rarely and under protest, but that was not surprising. The creature was a fish. (From *The Immense Journey* by Loren Eiseley with the author's permission. Published by Random House, Inc., 1957.)

The "snout" not only had external nostrils, but sturdy muscular fins, and lungs. Lungs and fish do not seem to go together today, but in Paleozoic fishes, the combination was commonplace. We assume that the ancestors of air-breathing fishes at first simply gulped air and then either regurgitated it, or passed it on through the digestive tract. Swallowing air in this manner is not a very satisfactory method of obtaining oxygen. Studies of living choanichthyans suggest that saclike bodies developed on the ventral side of the esophagus and became lined with a delicate network of tiny blood vessels for oxygen removal. These rudimentary lungs enlarged and gradually improved in the efficiency with which they extracted oxygen. In most modern fishes, the respiratory function of this organ has been lost, and the lung has been converted to the swim bladder which functions as an aid to hydrostatic balance.

Two tribes of hardy air gulpers lived during the Devonian. They are formally designated the Dipnoi and Crossopterygii. The first of these is not considered on the direct track toward tetrapods. However, the Devonian dipnoans may have been the earliest fish to solve the respiration crisis. *Dipterus* of Devonian age was a rather specialized fusiform fish with distinctive dentition for crushing small-shelled organisms and plants. These flattened teeth arranged in the floor and palate of the mouth are the distinguishing characteristics of dipnoans. The skeleton was partially replaced by cartilage and the paired fins were slender and leaf shaped. Although fossil evidence is obscure, it appears that the skeleton within the fins consisted of a central axis with numerous short side branches. The dipnoans survive today. Their name, "dipnoi," means "double breathers" and is suggested by the fact that they can breathe either by means of gills or lungs. Today the surviving dipnoans inhabit freshwater bodies in tropical regions of Southern Hemisphere continents. Their presence only south of the equator may have significance to the concepts of continental drift and movements of plates of the earth's crust. The Australain lungfish *Epiceratodus* closely resembles a Triassic precursor. In the South American genus *Lepidosiren* and in *Protopterus* of Africa, the paired fins are reduced to mere whiplike appendages (fig. 9.26). At the beginning of the dry season,

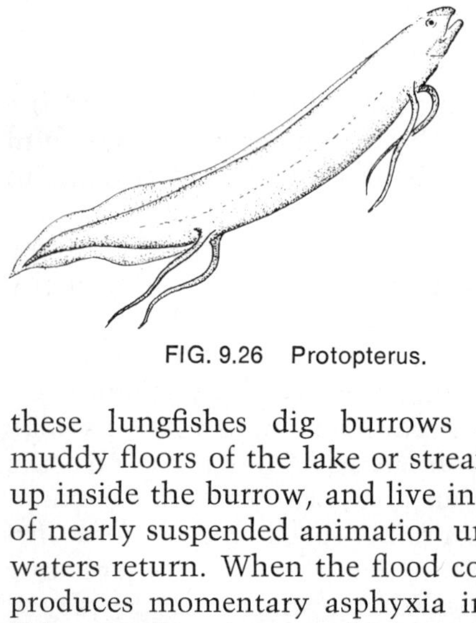

FIG. 9.26 Protopterus.

these lungfishes dig burrows in the muddy floors of the lake or stream, curl up inside the burrow, and live in a state of nearly suspended animation until the waters return. When the flood comes, it produces momentary asphyxia in these fishes resulting in convulsive movements toward the water's surface in order to obtain a lifesaving gulp of air. Odd that a fish could drown! Yet, Paleozoic strata in Texas have been found which contain the fossil remains of lungfish in their burrows. Apparently these fishes failed to escape their muddy tombs.

The crossopterygians are the true lobe-finned fishes. They are the ancestors of the tetrapods. *Osteolepis* (fig. 9.27) is perhaps one of the better known of the crossopterygians. In this genus, the paired fins had muscular lobes near their attachment to the body. Internally, a single basal element, the *humerus* (or femur for the pelvic area), articulates with the appropriate girdle and is followed by two bones, the *tibia* and *fibula* (fig. 9.28) for the posterior fins and the *ulna* and *radius* for the pectoral fins. As suggested by the passage from Eiseley, some crossopetrygians may have been able to walk for short distances over mudbanks. However, opinion varies on this point. Some paleontologists feel the animals may not have been able to withstand the force of gravity and body weight on their vital organs, or that the fins and strength were not sufficient. Others note the often amazing stamina of some fishes such as Florida's "walking catfish" (*Clarias batrachus*) which makes frequent and successful overland excursions. The important factor, however, is not whether the lobe-fins walked, but that they possessed the basic tetrapod limb plan. This plan persists in man and bird, frog and lizard, and has been cited as evidence for the validity of evolutionary theory for decades.

Crossopterygians differ generally from dipnoans in their higher degree of ossification and bony functional perfection. There is no trend toward substitution of cartilage for bone. The larger teeth (fig. 9.29) have infolded enamel similar to

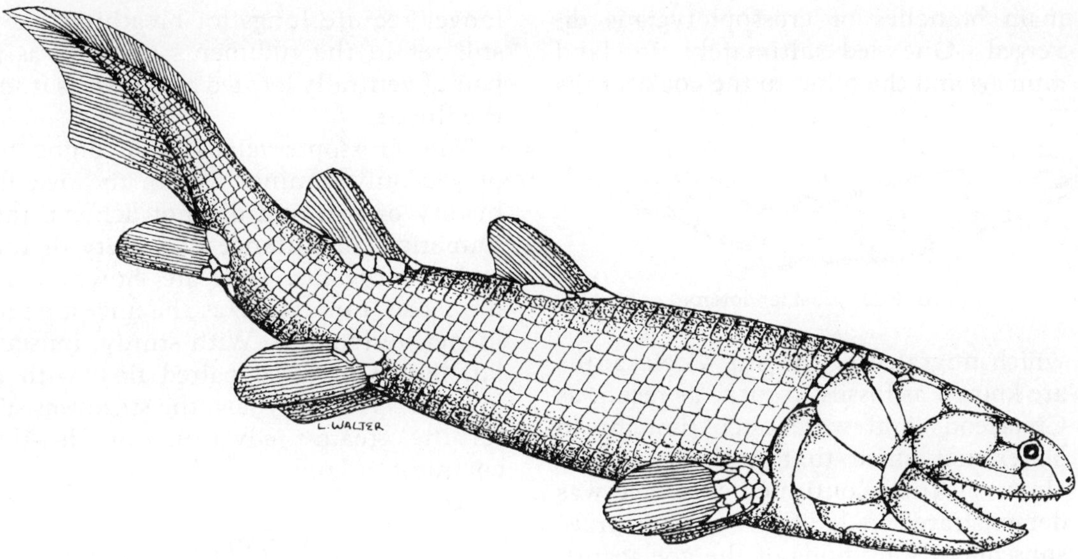

FIG. 9.27 Osteolepis.

that in the teeth of early amphibians, and the palatal teeth show the same type of alternate replacement. Furthermore, the bones of the skull and jaws are directly comparable to those of Late Paleozoic amphibians. Devonian crossopterygians, as indicated by the well-developed internal nostrils located in the ventral wall of the snout, had a well-developed sense of smell which they used to track down their victims. *Eusthenopteron* (fig. 9.30) was a crossopterygian which typifies the line that led toward the amphibians. The skull and vertebral column were prophetic of early tetrapods, and the tail assumed a distinctive three-pronged ap

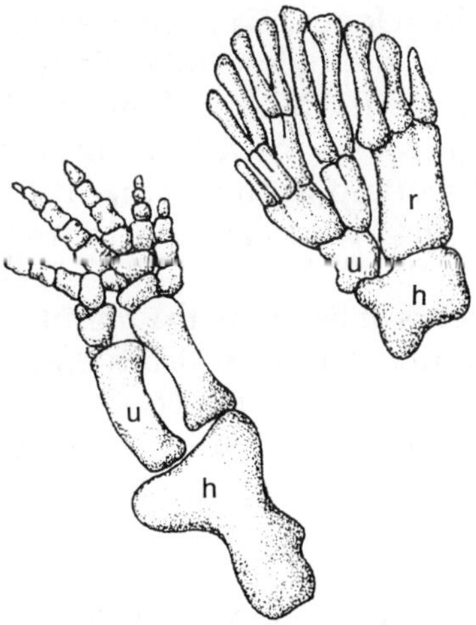

FIG. 9.28 *Diagrammatic drawings comparing the limb bones of a crossopterygian (upper right) and an early amphibian. (Modified from A. S. Romer, 1966* Vertebrate Paleontology, *3d ed., University of Chicago Press, p. 86.)*

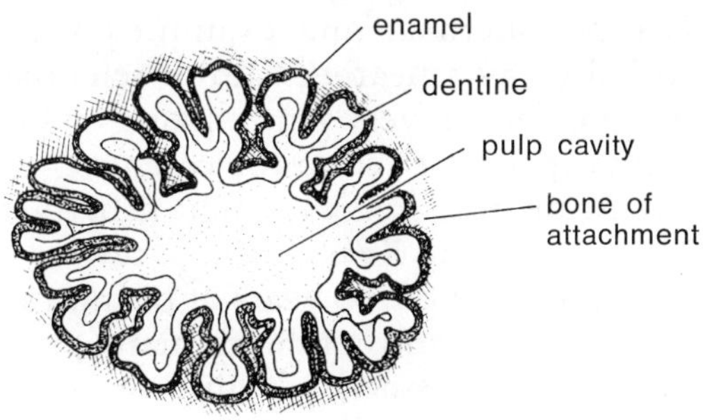

FIG. 9.29 *Cross section of a crossopterygian tooth showing infolded enamel.*

pearance. During the Devonian, two main branches of crossopterygians diverged. One led ultimately to land animals and the other to the coelacanths

FIG. 9.30 Eusthenopteron.

which migrated to the sea. Coelacanths are known as fossils in rocks as recent as Cretaceous and were once thought to be extinct since that geologic period. However, their continued existence was demonstrated in 1938 and several times subsequently by finds of the coelacanth *Latimeria* in deep water off Madagascar. As recently as 1972, a living specimen (fig. 9.31) was captured by a Comoro

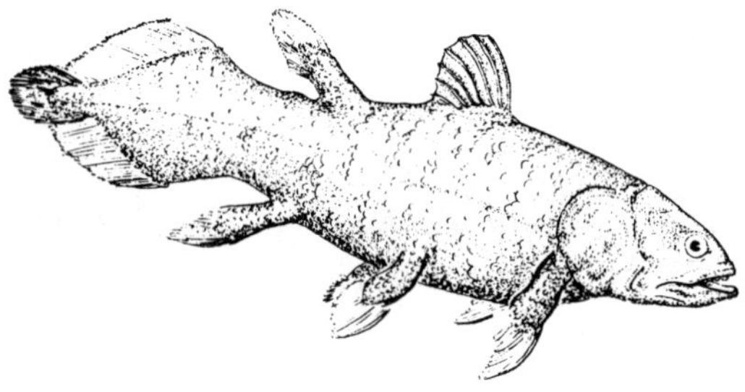

FIG. 9.31 Latimeria.

Islands fisherman and examined while still alive by a nearby team of scientists hunting for this very same "living fossil." Newspapers referred to the catch as "the five figure fish," for it brought fisherman Madi Youssouf Kaar £10,000 reward—the equivalent of 100 years of his normal income. *Latimeria* combines many primitive features associated with an earlier history with specializations derived at a later time from their long existence in the sea. For example, although they no longer require lungs for breathing, they still retain the rudiments of lungs as a pair of ventrally located outgrowths from the throat.

The crossopterygians and dipnoans played only a minor role in the overall history of fish and did not achieve the numerical importance of variety of the actinopterygians. They are, however, of immense importance as the ancestors of higher vertebrates. With sturdy, muscular, bone-supported paired fins, with a long and predatory jaw, the stage was set for the greatest adventure of all—the conquest of land.

SUMMARY

Aquatic vertebrates, loosely termed fishes, belong to at least five distinct classes; namely, the Agnatha, Acanthodii, Placodermi, Chondrichthyes, and Osteichthyes. Enigmatic toothlike microfossils called conodonts, which first appear in Upper Cambrian strata, may represent the oldest indications of vertebrate animals. More definite evidence of fishes is found in Ordovician rocks. It is still uncertain as to whether fishes originated in the oceans or in freshwater bodies, but the association of these early fragments of fish with the marine mineral glauconite, suggests fishes began their early evolution in the seas.

The oldest fishes are jawless members of the Class Agnatha. They are called ostracoderms. One of the better known of the ostracoderms is *Hemiclaspis*, a creature with the heavy bony armor that is the hallmark of this group. Ostracoderms were numerous during the Silurian and early part of the Devonian, but underwent extinction by the close of the Devonian. Their distant and specialized relatives, the lamprey and hagfish, are surviving members of the Class Agnatha.

One possible reason that may have contributed to the decline of the ostracoderms was the ascendency of jawed fishes with their greater adaptive possibilities. In brief, jaws evolved from one of the pairs of cartilaginous or bony processes (gill arches) that support the gills. In the first jawed fishes, the acanthodians, the upper jaw was attached to the skull by ligaments, and a full gill slit was present behind the jaw. In higher fishes, the upper jaw became anchored to the skull by the upper part of the first gill arch that had been located behind the primitive jaw. In the process, the first full gill slit was reduced to a small hole called the spiracle.

The placoderms were another group of primitive jawed fishes that were more numerous and varied than the acanthodians. The group included swift, streamlined swimmers like *Coccosteus,* some that resembled rayfish like *Gemuendina,* and heavily armored bottom dwellers such as *Bothriolepis.*

The higher fishes include the Chondrichthyes (sharks) and Osteichthyes (bony fishes). Both groups appear in the Devonian, and have persisted into modern times. As implied by their name, the skeletal parts of chondrichthyids are mostly composed of cartilage. The teeth and certain spines are exceptions and are the parts usually found as fossils. In terms of diversity and abundance, the Osteichthyes have been even more successful than their cartilaginous cousins. They may have evolved independently from placoderm ancestors, after which they diverged into two main evolutionary lineages. One of these comprises the familiar ray-finned fishes (Actinopterygii), and the other includes air-breathing fishes Choanichthyes).

The Choanichthyes can be divided into two orders; the Dipnoa, or lungfish, and the Crossopterygii, or lobed-finned fishes. The latter order is of major evolutionary significance because it included the ancestors of the first terrestrial vertebrates.

SUPPLEMENTARY READINGS AND REFERENCES

Black, Rhona M. 1970. *The elements of paleontology.* London: Cambridge University Press.

Colbert, Edwin H. 1969. *Evolution of the vertebrates.* 2nd ed. New York: Wiley.

Courtenay, W. R., Jr. 1970. Florida's walking catfish. *Ward's Bull.* 10 (69): 1–4.

Denison, R. H. 1956. A review of the habitat of the earliest vertebrates. *Fieldiana Geology* 2: 359–457.

———. 1967. Ordovician vertebrates from the western United States, *Fieldiana Geology* 16: 131–192.

Eiseley, Loren. 1959. *The immense journey.* New York: Vintage Books.

Koucky, F. L.; Cygan, N. E.; and Rhodes, F. H. T. 1961. Conodonts from the eastern flank of the Bighorn mountains, Wyoming. *Jour. Paleontology* 35: 877–879.

Millot, Jacques. 1955. The coelacanth. *Scientific American* 193 (6): 34–40.

Moy-Thomas, J. A. 1971. *Paleozoic fishes.* 2nd ed. London: Chapman & Hall Ltd.

Olson, Everett. 1971. *Vertebrate paleozoology.* New York: Wiley.

Ritchie, Alexander. 1968. New evidence on *Jamoytius kerwoodi white,* an important ostracoderm from the Silurian of Lanarkshire, Scotland. *Paleontology* 11 (1): 21–39.

Romer, Alfred S. 1959. *The vertebrate story.* Chicago, Ill.: University of Chicago Press.

Smith, Homer W. 1963. *From fish to philosopher.* New York: Doubleday.

Spoczynska, J. O. I. 1971. *Fossils.* Totowa, N.J.: Rowman and Littlefield.

Watson, P. M. S. 1951. *Paleontology and modern biology.* New Haven, Conn.: Yale University Press.

Yapp, W. B. 1965. *Vertebrates.* New York: Oxford University Press.

10 Two Ways of Life: Amphibians

We were amphibians, scaled and tailed
  And drab as a dead man's hand;
We coiled at ease 'neath the dripping trees
  Or trailed through the mud and sand.
Croaking and blind, with our three clawed feet*
  Writing a language dumb,
With never a spark in the empty dark
  To hint at a life to come.

—Langdon Smith, *Evolution*

REMODELING FOR LIFE ASHORE

It took tens of millions of years and many anatomical transformations to convert the hardy air-gulping fish into an animal that could comfortably spend a portion of its time away from water. The transformation was not complete, for amphibians continued to lay fishlike naked eggs in water or damp places. From this amphibian caviar sprang fishlike larvae which, like fish, used (and still use) gills for respiration. Later in their development, the endocrinal control exerted by such glands as the pituitary and thyroid, stimulated their metamorphosis into quadrupedal terrestrial adults.

The evolution of amphibians was not predestined. The evolution of limb and lung did not occur because fishes miraculously developed a yen for life on land. For a fish, land is a terrifyingly hostile environment. The dangers of dessication, the perpetual tug of gravity, the unfamiliarity or scarcity of food, and the daily change in temperature would cause any "sensible" fish to recoil at the thought of leaving the water home for which he was so well adjusted. For these reasons, it would seem more logical (although paradoxical) that the limb and lung did not evolve so that fish could explore the land, but rather were adaptations for finding and remaining in water during periods of drought. The "snout" laboriously dragged his body across the mud, not by choice, but for survival.

A host of changes accompanied the shift into part-time land dwelling. As the lung began to supersede the gill, blood had to be rerouted and the heart modified. A three-chambered heart (two auricles in parallel, and one ventricle) developed as a replacement for the dipnoan heart which had only a single

* Poetic license! Amphibians are basically five-toed.

auricle partly divided by a partition. For more efficient breathing, a pair of well-defined nasal passages led to the throat.

In order to better support the weight of the body, the pectoral and pelvic structures were greatly strengthened (fig. 10.1). the changes that transformed the spinal column into a sturdy yet flexible bridge of interlocking flanges, spines, and associated muscle attachments for support of the body. Indeed, the pattern and shape of vertebral elements provide the basis

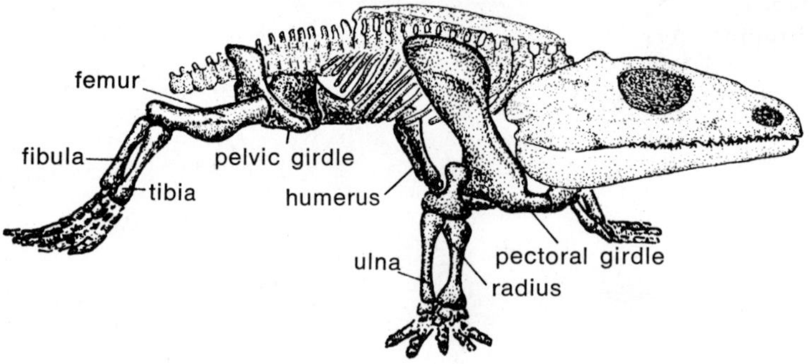

FIG. 10.1 *Partial skeleton of* Cacops, *an Early Permian labyrinthodont amphibian, emphasizing girdles and limbs.*

The fin supports of fishes are relatively simple structures consisting in part of surficial and internal supports. In the evolving amphibians, the internal bones are enlarged and improved for the attachment of powerful limb muscles. The pectoral girdle of early amphibians took the form of a broad U, and the pelvic structure resembled a V and was attached to the backbone by a single vertebra. As the initial step toward the evolution of what is loosely called a neck, the pectoral girdle was freed from its attachment to the skull, thus permitting movement of the head independently of the body. The most interesting development of all were for the classification of ancient amphibians. The vertebrae of crossopterygians had already become well ossified and were composed of four elements. A dorsal element enclosed the spinal cord and is thus termed the neural arch. On the ventral side, the intercentrum enclosed the notochord. Two smaller elements, the pleuracentra rested laterally between the other units. The earliest amphibians are called ichthyostegans (fig. 10.2), and their "backbones" are clearly similar to that of the Crossopterygians. The embolomeres were mostly water-dwelling forms with intercentra and pleuracentra suporting interlocking elements.

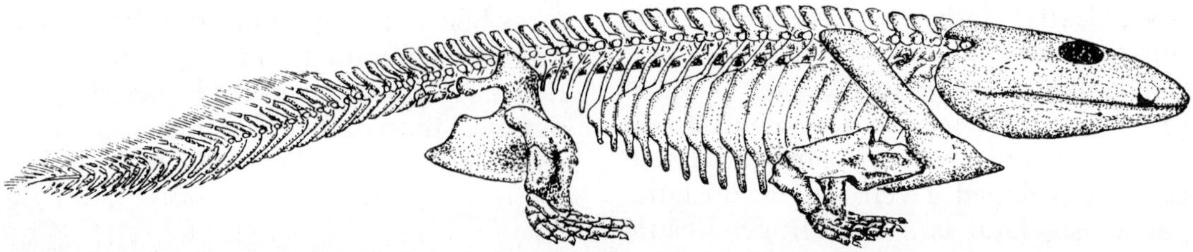

FIG. 10.2 Ichthyostega. *(Adapted from E. Jarvik, 1955, "The Oldest Known Tetrapods and Their Forerunners,"* Scientific Monthly *80: 142.)*

Many believe the embolomere lineage led ultimately to the reptiles. In a trend toward greater strength, the more terrestrial rhachitomes (fig. 10.3) had wedge-

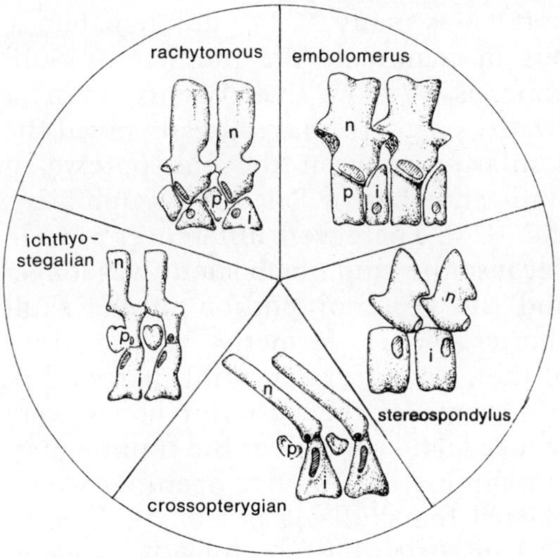

FIG. 10.3 *Crossopterygian, and four types of amphibian vertebrae;* n, *neural arch;* i, *intercentrum;* p, *pleurocentrum. (Modified from J. M. Weller, 1969,* The Course of Evolution, *McGraw-Hill, p. 410; and J. R. Beerbower, 1968,* Search for the Past, *Prentice-Hall, p. 450.)*

shaped intercentra, blocklike pleuracentra, and large, sturdy neural arches. The amphibians known as stereospondyls were aquatic forms that reversed the landward trend of their rhachitomous ancestors. Because they required less support, the intercentra were modified to spoollike elements and the pleuracentra lost.

In addition to changes in the supportive skeleton of early amphibians, there were significant changes in the sense organs. Fish, for example, do not have a device for transmitting sounds to the inner ear. As their medium is fluid rather than gaseous, they are more dependent upon vibrations which they receive along their lateral line sensory receptors. The problem of detecting sound in air required a new mechanism. The old hyomandibular which had originally been transformed from a part of the gill arch and subsequently put to use as a prop to hold the jaw and braincase together, was once again remodeled. Greatly reduced, it was transformed into a small sound transmitter called the stapes. On either side of the posterior portion of the skulls of Early Paleozoic amphibians, one can usually find a prominent notch. This notch, termed the tympanic notch, was for the accommodation of the tympanic membrane (eardrum). To complete the auditory apparatus, the old spiracle became the eustachian tube and middle ear. Sound waves impinging on the tympanic membrane were transmitted to the stapes and thence to the side of the braincase where the inner ear was located.

The eyes also required modification, for they were no longer continually immersed in water. Eyelids, glands, and ducts were developed to lubricate and to clean the eyes, and to protect them from drying, foreign objects, and the glare of unfiltered light. The "third eye," or pineal opening, seen even in the Early Paleozoic ostracoderms, is retained in the ancient amphibians. It is possible that this organ had a nerve channel leading to the brain and may have functioned in light perception. A working pineal organ can today be examined in the lizardlike tuatara of New Zealand.

Accompanying the increased reliance on the sense of vision were modifications for an improved sense of smell. The strange, flattened skulls of Paleozoic amphibians have greater distance between the eye orbits and the nasal openings than do crossopterygian fossil skulls. The openings for the nostrils continue through the palate into the mouth. Clearly these animals breathed and "smelled" through their noses. The increased reliance on smell may have had far-reaching effects. Behavior, for example, was undoubtedly strongly influenced by the olfactory sense. Because of the importance of the smell apparatus, and perhaps also the more advantageous anterior position of the part of the brain concerned with smell, the association centers of the brain which formerly occupied more posterior positions, shifted forward into closer contact with the olfactory lobes. From such inauspicious beginnings there would evolve—at a much later time—the unique machine for complex thought found in the expanded cerebral hemispheres of man.

Amphians have need of a body covering different from that of fishes. The fossil record shows a general loss of scales except along the belly where protection from the abrasive soil was needed. Skin is only rarely fossilized, but we may assume that the early amphibians had a moist skin, resistant to drying, and possibly capable of at least partial exchange of water, gases, and salt solutions. By means of retention and accumulation of dead skin cells, the skin became very thick in some early amphibians.

One of the more persistent trends in the overall evolution of vertebrates is the reduction in the number of individual segments in the skull. In fish skulls there may be as many as 140 different bones, but in mammals the number of skull bones is no more than twenty-seven or twenty-eight. We have already noted the similarity between the crossopterygian skull and that of Paleozoic amphibians (fig. 10.4). There were differences as well. Because the amphibian snout was longer and the posterior portion of the skull shorter, bones formerly at the front of the crossopterygian skull increased in size. Some of the posterior bones were reduced, and some near the trailing margin eliminated. The five openings which pierced the skull are obvious hallmarks of Late Paleozoic amphibians. Two of these were, of course, the nasal openings, two were for the eyes, and a fifth for the pineal organ. They provide spatial reference points for the identification of the surrounding bones.

DESCENDENTS OF THE SNOUT

The geologic record for the amphibians begins with the group called ichthyostegids. The original fossils had been ob-

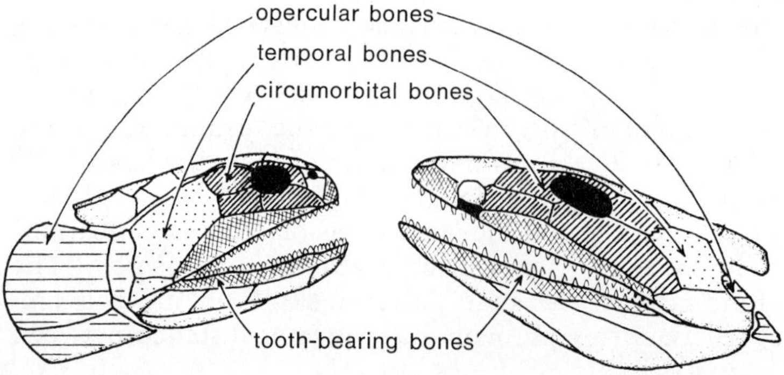

FIG. 10.4 *Comparison of skulls and lower jaws of crossopterygian fish (left), and the Devonian amphibian* Ichthyostega. *(Adapted from E. H. Colbert, 1969,* Evolution of the Vertebrates, *2d ed., Wiley, p. 77.)*

tained from freshwater Upper Devonian strata of Greenland in 1931. The skull had many similarities to that of the lobe-finned *Eusthenopteron,* and the body was somewhat laterally compressed with a fringe of fin around the tail. Unlike *Eusthenopteron,* however, *Ichthyostega* had true limbs with five webbed toes on each foot.

The amphibians which followed the ichthyostegids fall mostly within a group loosely termed "labyrinthodonts." They take their name from the intricately grooved and infolded structure of the teeth. During the coal age, labyrinthodonts wallowed in swamps and along streams eating insects, fish, possibly plants, and probably one another. These labyrinthodonts, known as embolomeres, had rather deep skulls as well as the sturdy *Eusthenopteron*-like backbone described earlier. Their limbs appear to have been somewhat less powerful than in some of the ichthyostegids, and this may reflect their preference for the aquatic habitat. The joint at the back of the skull (occipital condyle) was single and the braincase completely ossified.

Perhaps best known of the labyrinthodonts were the rhachitomes. The stout limbs of this group clearly indicate they were well adapted for life on land. Most were bulky, inelegant creatures resembling what might be described as a discarded blueprint for an alligator. The backbone had great strength with prominent zygapophyses for interlocking articulation between vertebrae. Dull-witted *Eryops* (fig. 10.5) with its flat head and ponderous body exemplifies the rhachitomes. They declined during the Permian and only a few forms survived into the Triassic.

Adaptations toward a terrestrial existence are reversed in the stereospondyls, the characteristic Triassic amphibians. Because they lived mostly in the water, gravity was less of a problem. In the more buoyant medium, larger size was more readily attained, and some stereospondyls became the largest amphibians ever to live. They tended to be flat in skull and body. The limbs and backbone became secondarily weak, and there was even a tendency toward substitution of cartilage for bone in parts of the skeleton.

The plagiosaurs were a group of bizarre and rather grotesque labyrinthodonts of the Late Triassic. These creatures had very broad flat skulls and diminutive legs.

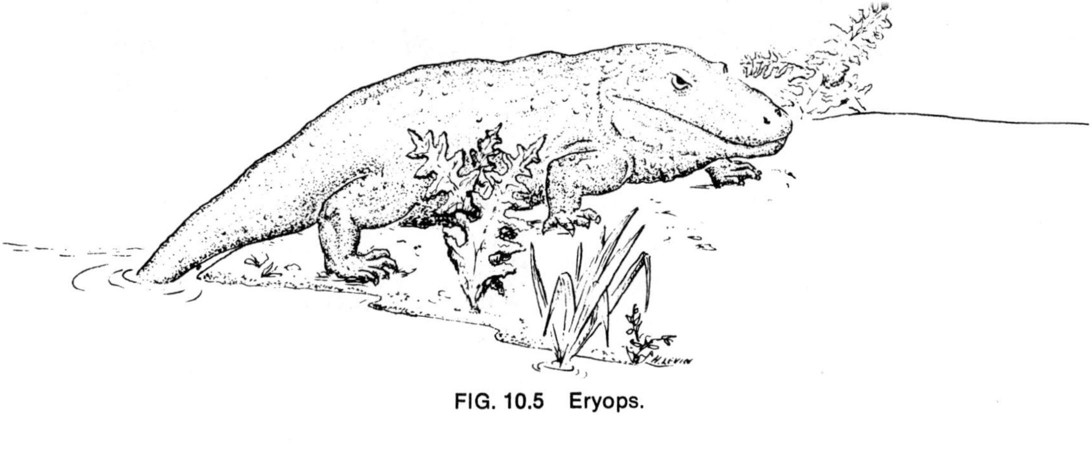

FIG. 10.5 Eryops.

Like the living salamanders, some may have developed external gills.

Lepospondyls were a group of Paleozoic amphibians that are not included in the labyrinthodont family. Unlike labyrinthodonts, which possessed vertebrae of several elements developmentally preformed in cartilage, the vertebrae of lepospondyls formed directly as spoollike bony cylinders around the notochord. They preferred certain habitats not dominated by the labyrinthodonts, and most lived a languid existence browsing in the aquatic undergrowth of swampy areas. Although not as numerous or diverse as the labyrinthodonts, they persisted from Mississippian into Permian times. *Diplocaulus,* with its peculiar boomerang-shaped skull, is usually the best remembered of the lepospondyls (fig. 10.6).

Surviving amphibians—frogs, toads, and salamanders—are placed together in the Subclass Lissamphibia. Toads and frogs had their beginnings in the Triassic, whereas the first salamanders are found in Cretaceous strata. One fossil salamander, discovered in Oligocene rocks in Switzerland, has become famous as a result of a small Latin volume written in 1726 by Johann J. Scheuchzer and entitled *Homo diluvii testis* (The man who is proof of the flood). Like many devout Christians of the early eighteenth century, the author believed fossils were relics buried by the flood described in the scriptures. The salamander was taken to be one of those sinful humans that perished by drowning.

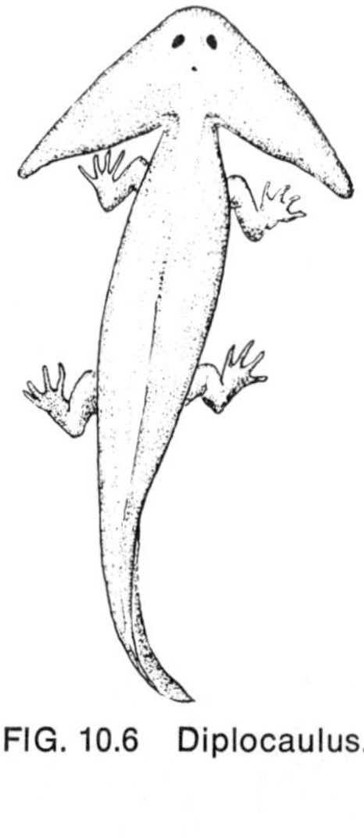

FIG. 10.6 Diplocaulus.

Today, amphibians are, in a sense, a defeated group. Their importance lies in the fact that they made the initial progress in the vertebrate exploitation of land. The total conquest of that environment, however, was to be the task of the reptiles.

SUMMARY

Late in the Devonian, we find the first fossils of terrestrial vertebrates—amphibians. It is clearly apparent from comparison of the pattern of skull bones that these animals evolved from lobe-finned (crossopterygian) fishes. The evolution of tetrapods from fishes involved many structural and physiological transformations. A three-chambered heart and improved olfactory abilities were among the changes that occurred. In addition, to oppose the pull of gravity on land, bones of the pectoral and pelvic girdles and limbs were enlarged and strengthened. Vertebrae assumed interlocking patterns to better support the weight of the animal. Ears, each with a tympanic membrane and sound transmitting stapes, evolved and were used to detect sound in the thin medium of air. No longer continuously bathed in water, the eyes were lubricated by special glands and ducts and protected from the sun's rays by eyelids.

The beginnings of amphian evolution are represented by a group of fishlike amphibians called ichthyostegids. From this general stock such extinct forms as anthracosaurs, stereospondyls, plagiosaurs, lepospondyls, and rhachitomes evolved. The clumsy looking rhachitomes provide the popular image of Paleozoic labyrinthodont amphibians. Surviving amphibians include the toads, frogs, and salamanders. Like their ancestors, these groups still are figuratively tied to the aquatic environment, for most deposit their eggs in water and have young that respire by means of gills.

SUPPLEMENTARY READINGS AND REFERENCES

Denison, R. H. 1956. The vertebrate ear. *Amer. Zool.* 6: 368–466.

Gregory, W. K. and Raven, H. C. 1941. Studies on the origin and early evolution of paired fins and limbs. *Annals N.Y. Acad. Sci.* vol. 42, 275 pp.

Noble, G. K. 1931. *The biology of the amphibia.* New York: McGraw-Hill.

Olson, E. C. et al. 1965. Evolution and relationships of the amphibia. *Amer. Zool.* 5: 263–334.

Westoll, T. S. 1943. The origin of the tetrapods. *Biol. Rev.* 18: 78–98.

11 The Reptilian Clan

Nevertheless, the thread of life ran onward, so that if you look closely you can see the singing reptile in the bird, or some ancient amphibian fondness for the ooze where the child wades in the mud.
—Loren Eiseley, *The Firmament of Time*

WAS IST DAS REPTIL?

What traits should an animal possess in order to qualify as a reptile? When forced to think about the question, one realizes that reptiles are recognized more by what they do not have, than by what they have. They are not amphibians because they lack aquatic larvae. Because they lack feathers, they cannot be considered birds. They are not mammals, for they are hairless. The traits reptiles do have include a dry scaly or cornified skin (in contrast to the smooth moist skin of amphibians). The reptilian heart is imperfectly four chambered. The "imperfection" arises from the fact that the two ventricles are not completely separated, except in members of the crocodile family. This is in contrast to birds and mammals in which the separation is complete. Finally, unlike amphibians generally, reptiles produce enclosed eggs rich in yolk. Because they are enclosed, these eggs are fertilized internally by copulation prior to being laid.

Unfortunately, these reptilian traits are of little use to the paleontologist who must seek definitive characters in skull, vertebrae, and limbs. Reptiles other than snakes typically carry their bodies well off the ground. As compared to labyrinthodonts, the limb bones are longer, more slender, and there is reduction in the number of bones in the wrist and ankle. The old pancake-flat head of Paleozoic amphibians is modified to a deeper form. The skull bones behind the parietals are reduced in number and size. The typical reptilian occipital condyle is a single knob. In the vertebral column, the pleuracentra are the largest elements, and the intercentra are reduced to small wedge-shaped units or lost altogether. Another difference can be found in the pelvic region. In amphibians there exists only a single sacral vertebra, whereas two or more are present in reptiles. Finally, one can discern in the reptilian skeleton a differentiation of the shapes of ribs in the abdominal, thoracic, and cervical areas, as befits an animal that is required to move about efficiently on land.

It is quite true that many of these reptilian traits were not clearly present in the earliest reptiles. The most ancient remains are Carboniferous in age and sadly inadequate in preservation. They are termed cotylosaurs. Less ancient strata of Early Permian rocks of Texas have yielded fairly complete cotylosaurian skeletons of the genus *Seymouria* (fig. 11.1). Here was an animal whose skull and dentition was not very different from some of the labyrinthodonts. However, the lacrimal bone at the orbit is pierced as if for a tear duct. Tears are of little use in water, and this bit of evidence favors the interpretation that seymourians were reptiles. There is other evidence as well, for careful study of seymourian remains indicate sexual dimorphism. There are, for example, apparent male and female differences in the distance between the first of the haemal arches that project below the tail vertebrae and the pelvis. Presumably, those specimens having a greater distance between the two points were females that required the extra space for emission of the large amniotic egg through the cloaca. The ultimate answer to the question of whether *Seymouria* was more reptile than amphibian would be best obtained from embryonic or larval evidence. In this regard, a form believed to be related to *Seymouria* has been discovered in larval form. It appears to have gills. Thus, a point is scored on the side of those believing in the amphibian ancestry of the seymourians. The debate, however, can be expected to continue for some time, as is to be expected when dealing with transitional fossils in one or another of the vertebrate classes. The interest they engender in students of evolution lies in the very fact that they are problematic mixtures of progressive and primitive traits. Indeed, the seymourians were precisely the kinds of transient animals one would expect to find along evolution's irregular, opportunistic, and experimental pathway.

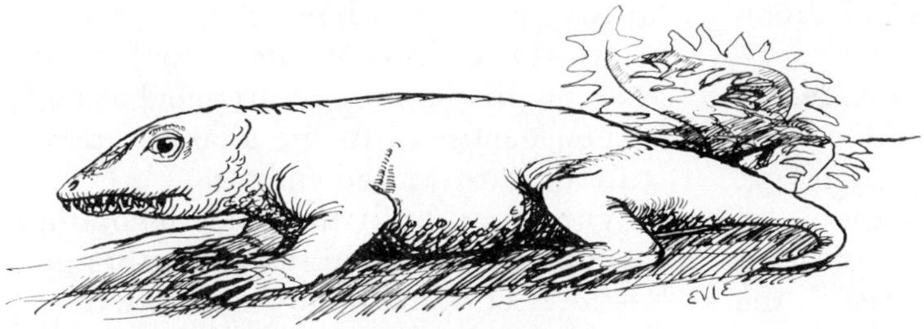

FIG. 11.1 Seymouria.

THE AMNIOTIC EGG

The ultimate link of the tetrapod to his ancient aquatic home was severed with the attainment of the amniotic egg. The advent of this new and prosaic object was a major milestone in the history of life. It provided the once aquatic larva with an encapsuled pond supplied with food, facilities for breathing, and a disposal unit for metabolic wastes. By analogy, the embryo had a relationship to its capsule reminiscent of the astronaut to his space laboratory. All essential life systems were "on board."

The amniotic egg can be characterized by its four essential features: a shell, albumen, extraembryonic membranes, and uric acid secretion. The shell in living reptiles is usually soft, with less calcium than in a bird's egg. Nevertheless, because of this covering, the egg must be fertilized in the female before it is laid. This operation is effected primitively by intromission of sperm into the cloaca without the aid of special organs of insertion. The function of the shell is primarily protection, but its porous structure does permit oxygen to pass into the shell and carbon dioxide outward.

Albumen is egg "white." It provides for the storage of water during embryonic development. Most reptiles lay their eggs in moist places where the external environment provides some protection against water loss. For this reason, the reptile's egg has less albumen than that of the familiar hen's egg as seen poached or boiled on our breakfast tables.

The third feature of the egg is the presence of four extraembryonic membranes. Clearly, these represent an advance from the single membrane or yolk sac of the amphibian. The membranes include, in addition to the yolk sac, the *amnion,* which contains a watery fluid in which the embryo is emersed throughout its development. The ancient watery habitat is thereby maintained. The *chorion* forms an outer sac which functions simply as a protective membrane. The chorion is highly vascularized and lies close to the inner surface of the shell in order to accomplish its respiratory role. The final membrane is the *allantois* which provides for respiration, absorption of albumen and water, and storage of excretory wastes.

The last important feature of the reptilian (and avian) egg is its method for nontoxic excretion. The metabolic wastes of animals are poisonous and must somehow be disposed of. In fish and amphibians, these poisonous substances are eliminated into the surrounding water which dilutes and renders them harmless. In mammals, the mother eliminates these wastes for the embryo. However, within the enclosed reptilian egg, direct elimination is not possible. As it is not possible to eliminate, the answer is conversion. In the embryos of advanced reptiles and birds, the kidneys have evolved the capacity to transform the lethal waste products into nontoxic uric acid, which is then deposited in the allantoic tissue without harm to the embryo.

When man built the space laboratory for encapsulating and sustaining life in a hostile environment, he did so with much knowledge of what would be required out there in space. To consider that Carboniferous vertebrates have similar foresight by preparing themselves for life on land, is of course, ridiculous. Logic dictates that the egg was the invention of animals that were still sloshing about in swamps and lakes. They had no particular "urge" for pioneering on the terrestrial frontiers. The vital orb developed, instead, in response to needs of the aquatic or semiaquatic amphibians, who—had they been able to dream—would not have dreamed of the egg's usefulness in later migration away from the shoreline.

The new kind of egg was for laying on land. What then might be its advantage for the amphibian that laid it there? The answer to the question seems readily apparent when one considers the prodigious number of eggs that are released and fertilized by living amphibians and the relatively few that attain maturity. Each season hosts of animals feast on this enormous gelatinous banquet of amphibian caviar. A gambler would speculate that in order to give the fertilized egg a better chance for life, the evolution of the new egg was sooner or later almost inevitably going to develop. The need of such a device is apparent even today among tropical frogs which appear to go to great lengths to avoid casting their eggs onto open waters. Some utilize damp depressions along the shore, or place them in depressions of logs or pockets in the dividing branches of trees. The "obstetrical toad" takes its name from the protective habit of wrapping the gelatinous strings of eggs about the legs of the male. Some frogs retain the fertilized eggs in the male's vocal pouch or within pockets in the female's skin. Infant survival, not a "desire" to become more completely terrestrial, is the reason for these devices —and for the evolution of the amniotic egg back in the sloughs of the Carboniferous.

HOLES IN THE HEAD—A MEANS OF CLASSIFICATION

Paleontologists still debate about the precise ancestry of the various branches of reptilian radiations, or whether there was a single or multiple amphibian stock. There is favorable opinion for the derivation of reptiles from embolomererian amphibians. There is also a consensus that reptiles then split into two distinct evolutionary lines, one leading to living reptiles and birds, and the other to mammallike reptiles and eventually mammals.

Because there are questions still remaining about the phylogenetic pathways of the various reptilian lineages, it has been expedient to classify reptiles according to the position and number of temporal openings in the skull. Some reptiles had two such "holes in the head" on either side of the skull. Others had one on each side, and still others lacked this feature altogether.

Most of us are more familiar with the appearance of mammal skulls than with skulls of reptiles. The large braincase seems the obviously noticeable part of such skulls. Reptilian skulls differ, and may be thought of as consisting of two containers. One of these is very small and holds the brain. It is located within and to the rear of the larger outer container. The earliest reptiles, the cotylosaurians, had a solidly roofed outer container. They, along with living turtles, comprise the Anapsida. The term "apse" is used for the bony arch between openings in the skull. Cotylosaurians had no holes, and correspondingly no "apses"—hence, the name Anapsida, "without arches" (fig. 11.2).

In anapsids, the muscles which moved the jaws originated on the deep medial surface of the temporal bones. From this point, they extended through holes in the palate to points of insertion on the mandible (lower jaw bone). Later, with the development of heavier musculature, lateral openings evolved and temporal muscles were able to bulge outward. In some reptiles, two such openings appeared. These are the Diapsida. Lizards, snakes, and the extinct dinosaurs and flying reptiles are all diapsid reptiles.

Only a single temporal opening

evolved in other lineages of reptiles. If that opening occurs high on the skull so that the postorbital and squamosal bones form the lower rim, the reptiles are classified as Euryapsida. (Parapsid skulls re-

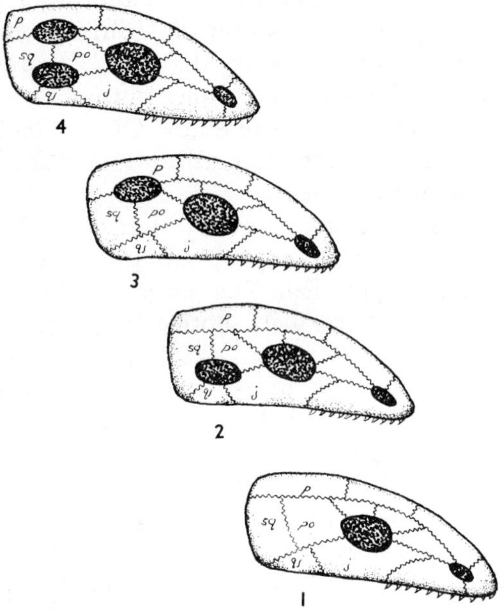

FIG. 11.2 *Reptile skull types; 1, anapsid; 2, synapsid; 3, euryapsid; 4, diapsid (*p, *parietal;* sq, *squamosal;* po, *postorbital;* j, *jugal;* qj, *quadratojugal).*

semble euryapsid, except that different bones occur beneath the temporal opening.) If, on the other hand, the single opening occurs low on the side of the skull and is bounded above by the squamosal and postorbital, then the reptiles are placed in the Synapsida. Synapsids are of particular interest because mammals apparently arose from this group.

BY LAND, SEA, AND AIR

Anapsids

The initial radiation of the reptiles began during the Pennsylvanian Period among the anapsids. By Permian time, primitive members of the other subclasses had also made their appearance. The group of anapsids known as cotylosaurs were the more common Pennsylvanian reptiles, and the cotylosaurian *Hylonomus* from Nova Scotia has the distinction of being the oldest known reptile.

Generally speaking, cotylosaurians were heavy-bodied animals with stocky legs. They include both carnivores and herbivores. Although cotylosaurs began their radiation in Early Pennsylvanian, they are better known from remains of Permian age. The genus *Limnoscelis* (fig. 11.3) from New Mexico is a satisfactory

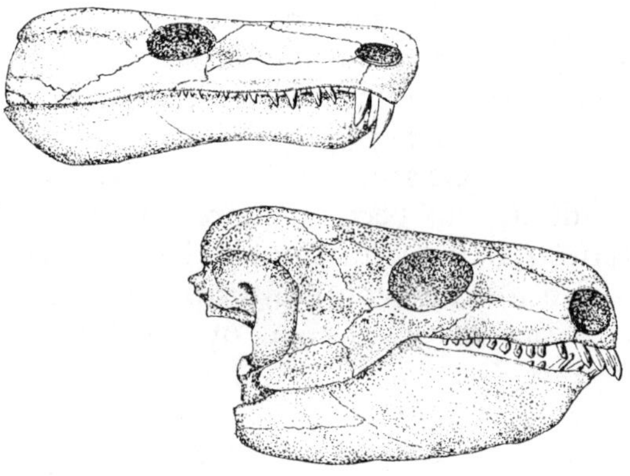

FIG. 11.3 *Skulls of two cotylosaurs,* Limnoscelis *(above) and* Diadectes *(below).*

representative. The solidly roofed anapsid skull of *Limnoscelis* retained the pineal opening of its amphibian precursors. Sharp teeth rimmed the margins of the jaws. Behind the head, the skeleton was decidedly reptilian in that the intercentra were reduced, the ilium expanded, and there were two sacral vertebrae. *Diadectes* (fig. 11.3), another cotylosaurian, presents a contrast in size and habit to *Limnoscelis*. This bulkier animal had peglike anterior teeth and broadened cheek teeth indicating a preference for plant food.

Yet another variation among fossil anapsids is seen in the aquatic mesosaurs (fig. 11.4) of Late Pennsylvanian and Early Permian time. Remains of mesosaurs are found in freshwater deposits on opposite sides of the South Atlantic in southern Africa and southern Brazil. They indicate the reptiles were two to three feet in length with long, slender, tooth-lined jaws. A deep tail and paddlelike feet complete the picture of a presumably aquatic or semiaquatic animal that inhabited freshwater bodies or coastal areas. Mesosaurians do not appear to have the high degree of specialization normally present in truly marine vertebrates. If these creatures were not the sort that could brave a journey across the expanse of the Atlantic Ocean, then their presence at the two widely separated localities may indicate connection or proximity of South Africa and southern Brazil near the end of the Paleozoic.

Synapsids

The early synapsids were Late Pennsylvanian contemporaries of many of the cotylosaurians. They have been named pelycosaurs and are of special interest as the general group from which the mammallike therapsids were to evolve. The Late Paleozoic pelycosaurs include the lizardlike ophiacodonts (fig. 11.5), sphen-

FIG. 11.5 Ophiacodon.

acodonts of predatory habit (fig. 11.6), and inoffensive plant-eating edaphosaurs (fig. 11.7). Edaphosaurs and sphenacodonts are favorites of school children who learn to recognize them from plastic models sold in department stores. The

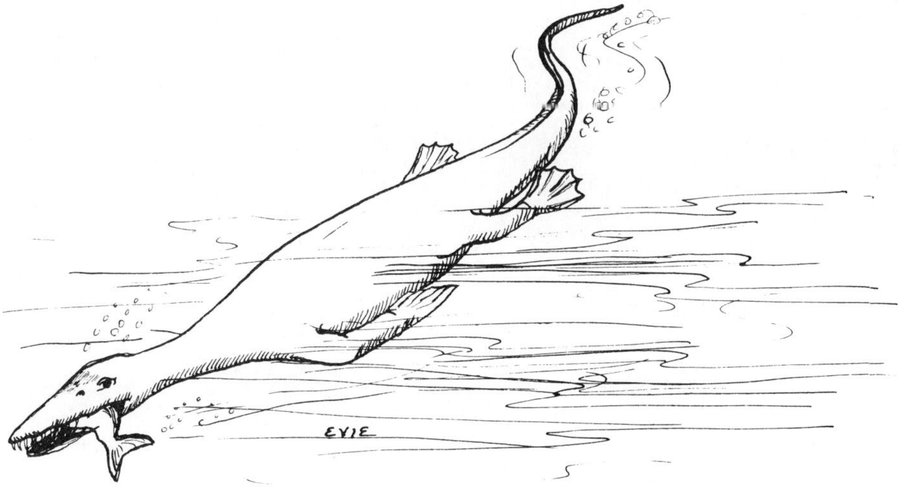

FIG. 11.4 Mesosaurus.

models vary in the faithfulness of their restorations, but clearly display the erect sail along the back that is the hallmark of the pelycosaurs. In life, the sail was supported by a series of greatly elongated dorsal spines attached to the vertebrae. The spines varied among different species. In *Edaphosaurus* they were heavier than in *Dimetrodon,* and were ornamented with short lateral bars. In other pelycosaurs, the spines were flattened so as to resemble a row of paddles. What could have been the function of these sail-fins?

One school of thought holds that the sails may have had a role in attracting sexual opposites. This seems unlikely, at least for species in which both males and females possessed sails. In this regard, it was once even seriously proposed that *Sphenacodon,* which lacked a sail, was really the male that sexually pursued the gloriously sailed *Dimetrodon,* and that these two forms were the same species. Remains of *Sphenacodon,* however, are found in New Mexico, and those of *Dimetrodon* in Texas. These localities were also separated by a Permian seaway. The need to overcome such geographic obstacles would certainly crush the ardor of even the most smitten pelycosaurian Lothario.

FIG. 11.7 Edaphosaurus.

There are still other theories. One suggests that sails were useful in frightening away enemies. Another proposes it was used in swatting pesky insects. However,

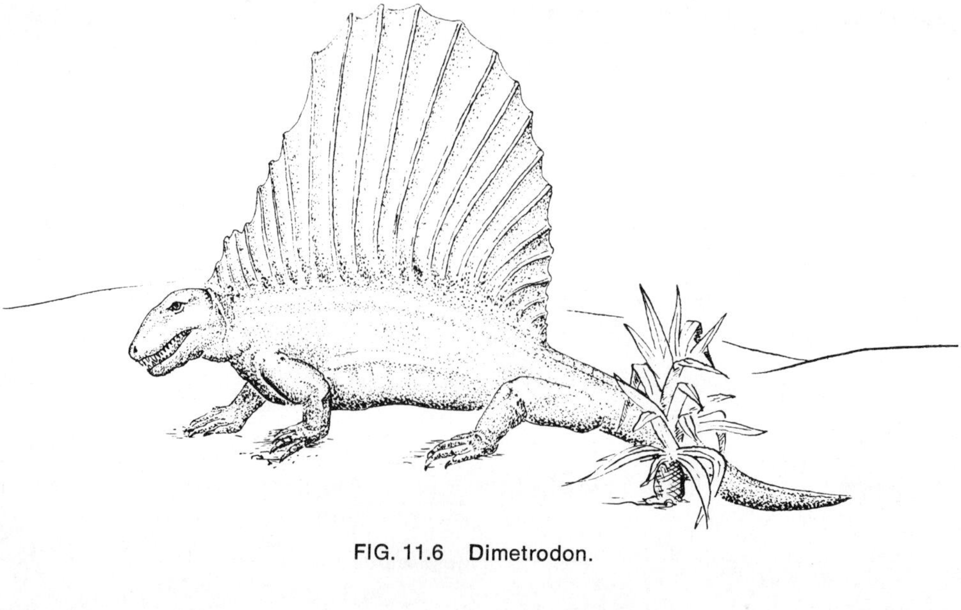

FIG. 11.6 Dimetrodon.

the most reasonable explanation is that the sail-fin served in body temperature regulation by sometimes acting as a heat receptor, and at other times a radiator. Pelycosaurs have traits suggesting they are the general stock from which mammallike reptiles and eventually mammals arose. It is interesting that they may have been attempting to achieve temperature regulation, that it was important to them, and that the need for such control later was to manifest itself in the system employed by mammals.

An examination of the habits of living reptiles lends validity to the temperature-regulation theory. Reptiles are regarded as "cold-blooded" because they lack the heat control mechanisms which mammals use to keep their body temperature uniform. When too cool, reptiles become sluggish and torpid. When heated more than a few degrees above 100°F., they may die of heat stroke. Desert reptiles avoid the lethal heat in various ways. Some confine their activity to early morning, night, or evening hours and retreat to crevices, bushes, or ravines for shade in the middle part of the day. One can imagine that pelycosaurs may have absorbed heat during the cooler parts of the day by placing themselves perpendicular to the incoming rays of the sun. When overheated, the sail could have been used to radiate excess body heat.

To man, the mammallike reptiles, or therapsids, are the synapsids of most interest. They were widely dispersed during the Middle to Late Permian and Triassic, and rich collecting grounds have been found in exposures of the Karoo deposits of South Africa.

Therapsids were predominantly small to moderately sized animals that evolved directly from the carnivorous pelycosaurs. Although they were a varied group, they all showed one or more features indicating their mammalian affinities. The skulls, for example, have a mammallike enlargement and lowering of the temporal opening. Closure of the pineal gland occurred, and the bones around the eye sockets were reduced in number. A double ball and socket articulation developed between the skull and the neck, and intercentra were eliminated between vertebrae. A clear indication of mammalian relationship is seen in the enlargement of the dentary bone at the expense of more posterior elements of the lower jaw. Teeth were firmly rooted into the jaw and, unlike more primitive reptiles, show differentiation into incisors, canines, and cheek teeth. This trend had begun, as a matter of fact, in the pelycosaurian ancestors of the therapsids. Oxygen consumption in mammallike reptiles can logically be assumed to be relatively greater than in ordinary reptiles. As a device to facilitate air intake and permit breathing even while food was held in the mouth, mammals have a separate nasal passage formed by growth of a secondary palate under the original roof of the mouth. This development correlates with the trend toward differentiation in dentition. Most flesh eating reptiles swallow their food whole, and then lie about slowly assimilating this difficult to digest mass. Once animals were provided with chewing and slicing teeth, food could be divided into small pieces, thus greatly increasing the total amount of surface area for attack by digestive juices.

Not all the changes in therapsids were in the skull and vertebrae. The limbs were swung into more or less vertical alignment beneath the body, and toe bones were reduced to the characteristic mammalian formula of 2–3–3–3–3. The way these animals walked was, of course, affected by these changes and can be visualized by comparing walking as seen in an ordinary dog and a salamander. In the latter, the gait is much like that in primitive quadrupedal reptiles. The upper bones of the limbs project sidewise like oars, leaving elbows and knees sticking out to the side. Sinuous bending of the body during walking helps to compensate for the lack of mobility in the limbs. Contrast this spraddle-legged gait with that seen in the dog. The spinal column of the dog is carried forward with a minimum of side to side movement, because the upper joints for the limbs (humerus and femur) are moved sideward and downward so that the limbs form vertical supports. Much of the body weight is taken off of muscles and placed on the bones, and locomotion becomes more efficient.

The beginnings of therapsid evolution can be traced to a group known as phthinosuchians. These animals, known principally from skull fragments, appear to be transitional between pelycosaurs and the early therapsids. The anomodonts represent one of the two major lineages that emanated from the phthinosuchians. Anomodonts include the dinocephalians, of which *Moschops* (fig. 11.8) is a well-known representative; the venyukoviamorphs, exemplified by *Venyukovia;* and the abundant and widely dispersed dicynodonts like *Dicynodon* (fig. 11.8).

The second major branch of therapsids was the more progressive theriodont lineage. The theriodont known as *Cynognathus* (fig. 11.9) is a representative form which embodied the mammalian trends

FIG. 11.8 *Two anomodont reptiles,* Moschops *(above) and* Dicynodon *(below).*

already described. *Cynognathus* was an active carnivore about the size of a small bear. The jaws carried a doglike set of teeth. Another form, *Gorgonopsia,* developed long fangs in the upper jaw presumably used for stabbing their prey in the same manner as did the saber-toothed cats millions of years later. The most advanced of the theriodonts are a group known as ictidosaurs. Ictidosaurs lived during the Triassic and Early to Middle Jurassic. In these theriodonts, the dentition was in an even more advanced stage of differentiation and the temporal opening quite large. Although the reptilian joint between the quadrate and articular bone persisted, a new mammallike joint between the squamosal and dentary bone developed. Ictidosaurs were about the size of rats or woodchucks and may have had rodentlike habits. They ideally bridge the gap between the therapsids and mammals. Some may have even possessed body hair.

FIG. 11.9 Cynognathus.

Although the advances made by therapsids were impressive, they were not sufficient to permit them to gain supremacy over the dinosaurs, which overshadowed them. As a matter of fact, most therapsids became extinct during the Triassic. A few survived, however, and evolved into mammals.

Diapsids

The diapsids have two temporal openings which are separated by the postorbital and squamosal bones (fig. 11.2). The group includes lizards, snakes, and crocodiles, as well as the extinct pterosaurs and dinosaurs. They ushered in the "age of dinosaurs" and flourished for a hundred million years. If numbers and variety attest to success in evolution, then the diapsids were indeed successful. Yet from the mammalian point of view, the reptilian success story may be unfortunate, for had this long period of reptilian supremacy not intervened, true mammals might have evolved one hundred million years earlier.

Diapsids are conventionally divided into relatively primitive "Lepidosaurs" (including lizards, snakes, rhynchocephalians, and eosuchians) and the more advanced "Archosaurs." During the Triassic, a group of archosaurians named thecodonts evolved and provided the stock from which dinosaurs arose. Thecodonts were lightly constructed reptiles that take their name from the fact that their teeth were set in sockets in the jaw bones—a characteristic of all archosaurs.

One of the obvious traits of thecodonts and many of their descendents was the tendency to rise up and walk upon their hind legs. Most were small, agile, reptiles up to three feet in length with longer and stouter hind legs and shortened forelimbs. *Hesperosuchus* (fig. 11.10) is a representative form. The forelimbs were probably used primarily for grasping and holding prey, while the long tail assisted in balancing the body so that the center of gravity occurred at the hips. When dashing after their prey, thecodonts probably poised body and tail horizontally. When resting, the hind legs and tail provided the stable support of a tripod.

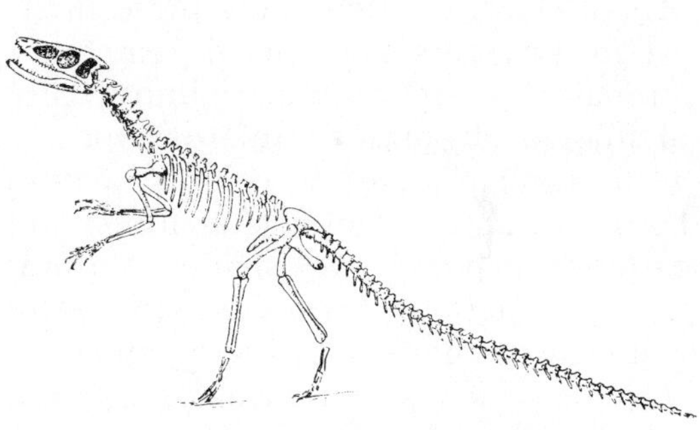

FIG. 11.10 Hesperosuchus.

Not all of the Triassic archosaurs were agile little sprinters like the thecodonts. The phytosaurs, for example, reverted to the quadruped stance and became crocodilelike in appearance (fig. 11.11, *Rutio-*

don). The most visible distinction between the Triassic phytosaurs and living crocodilians is in the position of the nostrils. In crocodiles, the nostrils are at the end of the snout, whereas phytosaurs had their nostrils just in front of the eyes. The Triassic "aetosaurs" were quadrupeds that were less like crocodiles. A representative form, *Desmatosuchus*, was characterized by heavy armor and spines.

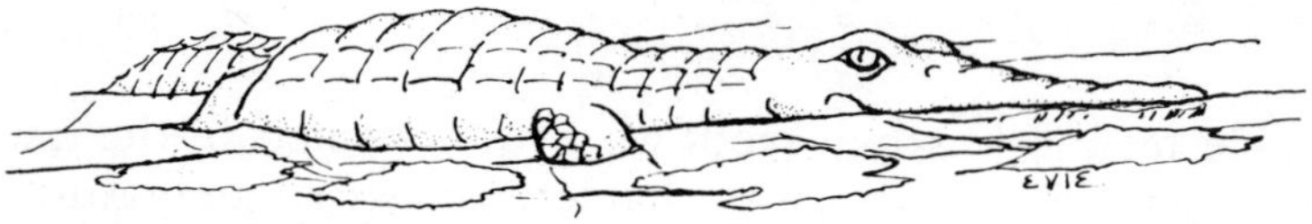

FIG. 11.11 Rutiodon.

Of all the reptiles living or extinct, none fascinate men so much as the dinosaurs. They are the most spectacular and familiar of all prehistoric animals. School children often show more interest in dinosaurs than in contemporary animals. We are reminded almost daily of dinosaurs by plastic miniatures for sale at department stores, in colorful picture books, in absurd Hollywood films, and cartoons in popular magazines. No museum of natural history seems really complete without a carefully mounted skeleton of a dinosaur. One forgets the jokes and cartoons in the cathedral atmosphere of the museum halls. Children and adults alike stare upward in wonder at the gaping silent skulls which seem to evoke riddles about the meaning and destiny of man.

We have noted the success with which the thecodonts developed the bipedal stance. Being two-legged animals ourselves, it is easy for us to see advantages in this development, especially in freeing the forelimbs from the tasks of support and locomotion. Bipedalism required, however, a strengthening and modification of the hind legs and hip structures. The arrangement of the bones in the hip region provides the criteria for classifying reptiles into two major groups; the Ornithischia with birdlike pelvic structure, and the Saurischia with reptilelike pelvis. In either case, the pelvic structure is composed of three pairs of bones fused together. Viewed from the side of a mounted skeleton one would see one bone of each pair or three in all. The uppermost of these bones is the ilium. It is firmly clamped to the backbone. The bone extending downward and slightly backward is the ischium. The remaining pubis joins the other two and projects downward and forward to help support the visceral mass. Where the three bones come together, there is a hollow area which affords a socket for the rounded end of the femur. In the saurischians, the arrangement of ilium, ischium, and pubis is triradiate. However, in the ornithischians, the pubis extends both forward and aft and the ischium is swung into close proximity (fig. 11.12). The rearward prong of the ornithischian pelvis is a diagnostic feature.

The first saurischians, like their thecodont ancestors, were small, lightly built bipeds. Their general characteristics are particularly well represented in the skeletons of *Coelophysis* (fig. 11.13) from the Triassic Chinle Formation of New Mexico. Long before men knew of dinosaurs, the tracks of *Coelophysis* were found as

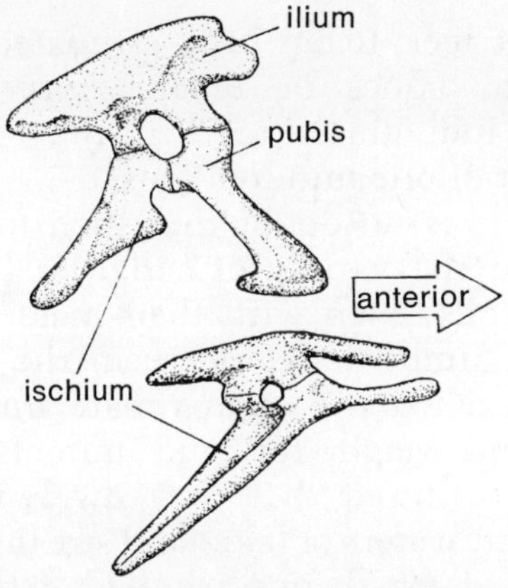

FIG. 11.12 *Pelvic structures in dinosaurs, Saurischian (above) and Ornithischian (below).*

impressions in sandstone and attributed to giant birds. Indeed, like birds, some of the bones in this lithe reptile were hollow. The hip socket, or acetabulum, was perforated rather than a hollowed notch. The delicately constructed skull was armed with sharp, compressed, and serrated teeth attesting to a preference for meat as a dietary mainstay. It is likely, that *Coelophysis* and his ilk represent the beginnings of the evolution of the large carnivorous bipeds of the later Mesozoic.

What then of the beginnings of the plant-eating sauropod lineage? The ancestry of this group is best exemplified by the so-called "pro-sauropod" *Plateosaurus* of the Late Triassic (fig. 11.14). More heavily boned than the more primitive theropods, this dinosaur was primarily herbivorous as indicated by the flattened spatulate teeth. The bones were no longer hollow and the legs and feet heavy. Even the forelimbs were larger, so that the animal could, when required, walk on all four legs. In short, *Plateosaurus* and related pro-sauropods initiated the trend toward giganticism that culminated in the enormous sauropods of the following Jurassic Period.

The close of the Triassic Period in geologic history was like the preceding Permian, a time of many extinctions. Labyrinthodonts, cotylosaurs, thecodonts, and all but a few therapsids disappeared. The stage was set for the advent of the dinosaurian giants.

The sauropods, or herbivorous branch of the Saurischia, appear at the beginning of the Jurassic. They were long-necked, long-tailed, exceedingly large reptiles that vividly suggest evolution from plateosaurians that reverted to the four-legged stance in order to support their tremendous bulk. There were several types of sauropods that lived in Africa,

FIG. 11.13 Coelophysis.

Eurasia, and America during the Jurassic and Cretaceous. They differed in details of anatomy that suggest early diversification. A well-known representative, to many people almost synonymous with the word "dinosaur," is *Brontosaurus* (more properly *Apatosaurus*), the "Thunder Beast" (fig. 11.15). Like many of the Jurassic ruling reptiles, its remains occur in the Morrison Formation of Colorado. This great beast was over sixty feet long and probably weighed in excess of thirty tons. *Diplodocus,* a Jurassic contemporary of *Brontosaurus* was the longest of the sauropods, one specimen measuring nearly eighty-eight feet from head to tip of tail. It was, however, more slender, and probably weighed only about a tnird as much as its bulkier cousin. In both *Brontosaurus* and *Diplodocus,* bipedal ancestry is revealed by their longer rear limbs. The exception to the rule was *Brachiosaurus* which had longer forelegs resulting in a shoulder height in excess of twenty feet. It has been estimated from isolated bones of *Brachiosaurus,* that some individuals reached the colossal weight of one hundred tons.

There is no doubt that such tonnage limited the variety of activities for the sauropods. Even with their massive pillarlike limbs, it would seem the giants had reached the approximate limits in size and weight for land animals. It is often surmised that they dwelt in the bouyant waters of lakes and streams, and overland excursions required extraordinary effort. The nostrils in sauropods were well suited for an aquatic animal. They were located high on the skull for suitable "snorkel action." Weight was concentrated over the hip region for greater stability while partly submerged. They probably led lives somewhat similar to the hippopotamus, browsing on aquatic plants and rarely venturing very far from water.

The support for the sauropod's avoir-

FIG. 11.14 Plateosaurus.

FIG. 11.15 Brontosaurus.

dupois consisted of four massive legs, terminating in typical graviportal (weight-carrying) feet. Toes were arranged in a spreading conical manner and a thick elastic pad probably cushioned the bottom of the foot. The front and rear feet had five toes, some of which bore claws. Vertebrae often numbered over eighty, were joined together by powerful ligaments and auxilliary bearing surfaces. The sauropod neck had a snakelike flexibility and terminated in an absurdly small head. The jaws were weak, and peglike teeth were unimpressive.

One cannot help but wonder how much food was consumed each day by a healthy sauropod. A not very good analogy is provided by the elephant which consumes over a hundred pounds of food a day. It takes the elephant ten to twelve hours to do this. For a sauropod, which was six times larger than an elephant, one would expect dinner never ended. Perhaps however, the task was not as prodigious as one might be led to expect from the size of sauropods. They were cold-blooded beasts, and thus needed less food for a given body weight than do the warm-blooded elephants. Then too, food was apparently abundant and readily accessible.

For sauropods, there were probably advantages in being big. Large size affords protection. The lion rarely attacks the elephant. It is also possible that giganticism provided partial control of body temperature. The ratio of surface area to mass for an animal decreases as size increases. Consequently, a large animal has proportionately less surface for heat loss. Like the large pot of hot water that takes longer to cool down than the small pot, body temperature would fluctuate more slowly.

The idea that sauropods were stupid, clumsy creatures also bears examination. Like the hippopotamus which appears obese and awkward on land and yet is a graceful swimmer, so also the sauropods were probably supple and graceful in their preferred habitat. True, the brain was small, but probably adequate for an animal that relied heavily on instinct. By comparison, many invertebrates have miniscule brains relative to body size, but are able to function in complicated and efficient ways.

Many questions about the habits of dinosaurs remain unanswered. One wonders about the manner in which they engaged in courtship and coitus, for example. As one stands in the museum hall of dinosaurs and views the ponderous bones, spines, and armor plate, one cannot but imagine the cooperation and ingenuity required. However accomplished, the 120 million year history of the dinosaurs indicates that coitus was accomplished efficiently. Once the female was pregnant, the course of events was probably not unlike that in many living reptiles. The female may have scooped out a hollow, laid her eggs in it, and covered the depression with sand. We know something of at least some dinosaur eggs. Eggs found in Mongolia were oblong and about eight inches long. They are generally believed to have come from relatively small beasts of perhaps eight feet in length. Extrapolating to the sixty foot sauropods, one might whimsically imagine an egg over five feet in diameter. Truly, it would make an omelet sufficient for several battalions. Of course, such an eggs is an absurdity. The fluid pressure would require internal supports or a shell so thick the baby reptiles would not be able to break out. Calculations indicate that, for eggs, a diameter of ten to thirteen inches would be about maximum.

With the abundance of herbivorous saurischians during the Jurassic and Cretaceous, there were a number of carnivorous forms. As is the case today, they performed an important role in achieving a balanced biotic community and were important factors in the operation of natural selection among herbivores. Also much like the situation today, a minority of carnivores fed upon a preponderance of herbivores. These carnivorous dinosaurs are termed theropods. Theropods were a varied group. Some were small and agile, like the lively *Ornitholestes* (fig. 11.16) which rarely exceeded six feet in length. *Allosaurus* of the Jurassic and *Tyrannosaurus* (fig. 11.17) of the Cretaceous are the familiar Mesozoic giants. The latter often attained lengths of over forty feet and weighed in excess of five tons. Theropod hindlimbs were stout and

FIG. 11.16 Ornitholestes.

FIG. 11.17 Tyrannosaurus.

muscular. Huge hooked claws for tearing flesh protruded from each of three toes. A fourth toe, more or less rudimentary, bore a smaller claw. The forelegs were much smaller and also had three clawed digits. In order to support the animal's weight, the vertebrae developed interlocking processes. This increased their ability to bear the load, but decreased the flexibility of the spinal column. It is likely that when on the move, theropods carried their bodies nearly parallel to the ground. When at rest, they probably had their bellys on the ground with neck and tail outstretched in either direction. Walking among theropods may have been much like that of an ostrich. Each foot was emplaced before the other was lifted. The stiff counter-balancing tail was probably held habitually off the ground.

The heads of the carnivores were large as befits an animal that must kill with its jaws and teeth. Large fenestra pierced the high, narrow skulls both in front and behind the eye orbits. Teeth, doubly serrated and up to six inches long, lined the powerful jaws. Little is known for sure about their habits. Chances are they stalked about until they came upon their prey, or lurked quietly in the shadows until an unfortunate victim passed their way. A quick rush, a clamping of the jaws into the prey, and much tearing and tugging with teeth and claws completed the unsophisticated kill. It is also possible that theropods were not so fastidious in their choice of a meal that they didn't enjoy an occasional repast of carrion. Many other questions about these ferocious-looking beasts will probably remain forever unsolved. Were they able to make vocal noises? If so, was it a hiss like a snake, or the sonorous grunt of an adult crocodile. Did they have territorial instincts as do many living mammals? To what degree did their tiny brains permit elaborate reflex patterns? What was their normal life span, their nutritional requirements, growth rate, longevity, and diseases? It is nearly three hundred years since the first theropod femur was discovered, and yet dinosaurs still provide fascinating subjects for paleontological speculation.

The other major dinosaur line, the ornithischian, makes its appearance near the end of the Triassic. Ornithischians were plant eaters. Teeth in the forward part of the jaws were lost, and a beak suitable for cropping vegetation replaced the lost teeth. A new bone, the predentary, became the lower element of the beak.

The oldest, fairly complete, ornithischian remains permit one to reconstruct a squat, stocky, quadrupedal, armored form. Reflecting their bipedal ancestry, the hindlimbs were somewhat larger than the forelimbs. These primitive armored reptiles were the ancestors of the oddly ornamented stegosaurs (fig. 11.18). Stegosaurians sported a double row of about twenty bony plates along their arched backs. Two pairs of sharp spines added protection to the tail. In the early days of dinosaur hunting, paleontologists argued over whether the back plates were paired or alternating, moveable or fixed, or lay flat or erect. Specimens discovered more recently indicate they were erect, immovable, and alternating. Each was attached to the underside of the skin by a broad base. No one knows for sure their purpose. They appear to offer little real protection. Yet, one can imagine *Stegosaurus* in ponderous, clacking flight from a hungry flesh eater who clamps his jaws instinctively into a very accessible but not vital back plate only to be suddenly gored by the vicious tail spikes.

Stegosaurians, like all the great reptiles, had a very small brain (2.6 ounces). It also had an enlarged opening in the sacrum which has prompted an entertaining but incorrect fable that a second "brain" existed there. Actually this space was needed to accommodate nerves from

FIG. 11.18 Stegosaurus.

the legs and viscera where they entered the spinal cord. It may, as suggested by H. M. Smith, have provided the space for a glycogen body. Thus, another delightful but fallacious story about an ancient creature, which like a hook and ladder fire truck had controls both fore and aft, must be put aside. The myth gained international fame as a result of the publication of the following poem by *Chicago Tribune* reporter B. L. Lester in 1912.

Behold the mighty dinosaur,
Famous in prehistoric lore,
Not only for his power and strength
But for his intellectual length.
You will observe by these remains
The creature had two sets of brains—
One in his head (the usual place),
The other in his spinal base.
Thus he could reason "A priori"
As well as "A posteriori."
No problem bothered him a bit
He made both head and tail of it.
So wise was he, so wise and solemn,
Each thought filled just a spinal column.
If one brain found the pressure strong
It passed a few ideas along.
If something slipped the forward mind
'Twas rescued by the one behind.
And if in error he was caught
He had a saving afterthought.
As he thought twice before he spoke
He had no judgement to revoke.
Thus he could speak without congestion
Upon both sides of every question.
Oh, gaze upon this model beast;
Defunct ten million years at least.

During the Cretaceous, stegosaurs were succeeded by even heavier reptiles that somewhat resembled giant armadillos. These bulky squat ankylosaurs, as they are called, had bony plates covering the length of their twenty-foot backsides. Short, sturdy spikes extended outward along the animal's flanks. A great ball of bone at the end of the tail in *Ankylosaurus* (fig. 11.19) could be used to bludgeon an unwary foe. In the design of ankylosaurs, nature seems to have provided a model for the tank long before there were wars or men to fight in them.

FIG. 11.19 Ankylosaurus.

More successful than either the stegosaurians or ankylosaurs were the oddly droll looking ornithopods. They are familiarly referred to as the "duckbills." Although they were bipedal animals, the forelimbs were never as drastically shortened as in the theropods. The two-legged stance was probably used while feeding; and, when required, they could revert to a quadrupedal gait. To the rear of the characteristic beak were ridged cheek teeth suitable for grinding up vegetation. Toward the end of the Cretaceous, a group of ornithopods termed hadrosaurs (stout reptiles) became particularly large and fairly abundant. *Camptosaurus* (fig. 11.20) was a typical hadrosaur with laterally flattened tail and webbed feet. The face was distinctive, with the forward part of the jaws flattened and spread to form a toothless bill. Behind this toothless part of the jaw, there was a tight mosaic of row upon row of grinding teeth—as many as two thousand in all. Worn teeth were replaced continuously. This type of dentition indicates a diet of coarse and abrasive plants and suggests that the hadrosaurs may have fed on silica rich *Calamites,* or even some of the emerging angiosperms.

Yet another peculiarity of the hadrosaurs was the development in some species of odd bony crests and tubular outgrowths of the nasal passages on the top of the skull (figs. 11.21 and 11.22). The nasal passages passed up these tubes and then doubled back to enter the pharynx. We can only surmise the purpose served by these bizarre structures. There is a good possibility that they provided additional mucous membrane tissue from which led nerves with an olfactory function. Unarmored and slow, the duckbills probably could well have used a superior sense of smell in the early detection of approaching carnivores. The instinctual response was very likely a quick plunge into the protecting waters of an adjacent lake or stream.

Not all of the hadrosaurs had skull crests, and some that had crests lacked the nasal tubes. One such group is represented by the Cretaceous genus *Pachycephalosaurus* (fig. 11.23). These were nature's original "bone heads." The skull in these forms consisted mostly of solid bone except for a tiny space to accommodate the brain. *Pachycephalosaurus* prob-

FIG. 11.20 Camptosaurus.

FIG. 11.21 Parasaurolophus.

FIG. 11.22 Corythosaurus.

ably used his battering ram head to protect himself, defend his territory, or fight for females.

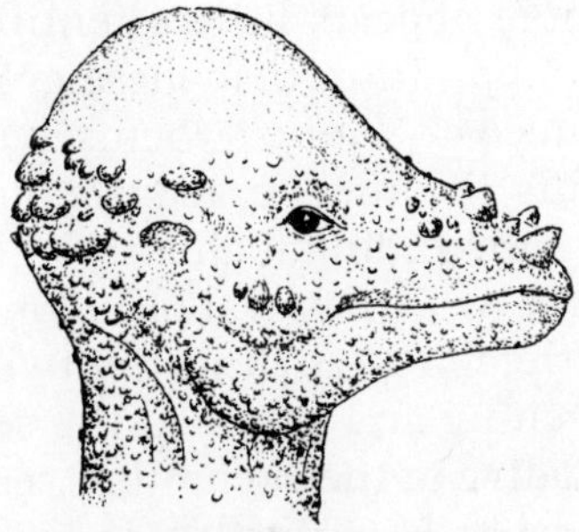

FIG. 11.23 Pachycephalosaurus.

The final group of ornithischians were the huge, quadrupedal, horned ceratopsians. Early forms lacked the horns from which the group takes its name. One of these, *Protoceratops* (fig. 11.24), has

FIG. 11.24 Protoceratops.

gained fame as the animal responsible for the fossilized dinosaur eggs discovered by the American expedition to Mongolia in the 1920s. Some of the fossilized eggs even contained the vestiges of embryos. The more familiar, later ceratopsians possessed a median horn just above the nostrils, and in some genera there was an additional pair which projected from the skull over the eyes. Typically, the head was very large in proportion to the body and was outfitted with a shieldlike posterior extension. The shield was probably used for protection as well as for attachment of powerful jaw and neck muscles. In some species, the bone of the shield had large openings that were covered in life by leathery skin. Additional horns and spines ornamented the periphery of some Cretaceous ceratopsians. All species had a characteristic parrotlike beak at the front of the jaw. Scarred shields and broken horns of ceratopsians are mute testimonials to their bellicosity. One can only imagine the lethal shock experienced by a hungry *Tyrannosaurus* in his instinctual rush into the shield and spike defense of a stalwart *Triceratops* (fig. 11.25).

"The flying diapsids" sounds like the billing for a group of circus acrobats. Nevertheless, it is an appropriate name for that group of diapsids known as Pterosauria which took to the air. They were animals which varied in size from small insect eaters only a few inches long, to mighty gliders with wingspans of several feet that fed upon fish, belemnites, or crustaceans. The typical pterosaur had a relatively large head. The eyes were protected by a ring of sclerotic plates. The jaws were long and lined with long, thin, forwardly slanted teeth. To provide for the high degree of coordination and nervous control required by flight, the brain was rather large for a reptile. In pterosaurs, the fourth finger was lengthened to support the wing. The next three fingers were of ordinary length and terminated in claws. The fifth finger was lost. The wing was a sail of skin stretched from this fourth digit to the body and rear limb. Shoulder and breast bones were modified to provide anchorage for the strong muscles needed to operate the wings. There were two general groups of pterosaurs. One of these, typified by *Rhamphorhynchus* (fig. 11.26) of the Jurassic, had long tails, whereas members of the second group had short tails. The Cretaceous form *Pterandon* (fig. 11.27) is one of the better known of the short-tailed pterosaurs. Although it had a body of the approximate size of a goose, its wingspan exceeded an astonishing twenty-five feet. The bone was very lightly constructed, and it is estimated that *Pterandon* only weighed about forty pounds. The animal probably soared like an albatross over the inshore areas of the Cretaceous seas snapping up various sea creatures in its toothless jaws. Using a standard aerodynamic formula, it appears that *Pterandon* could easily become airborne even from the crest of a wave, provided it could face into a wind of somewhat over fifteen miles per hour. Thus there appears little scientific proof for the supposition that these great kites were forced to launch themselves from high cliffs in order to acquire sufficient air speed for gliding flight.

Anyone who has walked along a deserted stretch of beach in the twilight of early evening and watched the sea birds soar and dive in their relentless search for food, has little difficulty in imagining

FIG. 11.25 Triceratops.

FIG. 11.26 Rhamphorhynchus.

FIG. 11.27 Pterandon.

a Cretaceous sunset populated with pterosaurs. They were successful aviators for millions of years, yet were unable to persist into the Cenozoic. By that time, the birds had begun to assert their supremacy as aerial navigators.

Having examined reptiles of the land and air, it is time to look at a group which contains a number of marine forms. Most, but not all, of the marine reptiles were euryapsids. Turtles, for example, are anapsids, whereas crocodilians and mosasaurs are diapsids. Thus we see that more than one order of reptiles returned some of its members to the sea. This is not an uncommon occurrence in evolution, for once any group becomes established in a particular environment, some of its descendents may begin to expand toward the niche perimeters and finally adapt to new environments.

The modifications for life in the sea were nearly as extensive as those that occurred when animals left the water for life on land. Euryapsids, as well as the diapsid crocodiles and mosasaurs, made their adjustments in various ways. Paddle-shaped limbs and streamlined bodies provided once again for efficient movement through the water. Unable to abandon lungs and convert back to gills, they evolved increasingly more efficient lungs. In some cases adaptations were so complete that the animals were unable to go ashore temporarily to lay their eggs. For these reptiles, reproductive adaptations permitted birth of young alive and while at sea.

Evolution of the euryapsid reptiles began in Permian with the appearance of small, lizardlike creatures called protorosaurs. In the succeeding Triassic Period, the nothosaurs arose from protorosaurian ancestors. The nothosaurs were early representatives of a rather diversified group known as sauropterygians. They had clearly initiated the modifications for marine life by developing long streamlined bodies; short limbs terminated in webbed toes; and a secondary palate separated the mouth from the air passages. The sauropterygians known as placodonts (fig. 11.28) were bulkier forms with distinctive pavement type teeth in the jaws and palate for crushing the shells of molluscs and other invertebrates. By far the best known of the sauropterygians were the plesiosaurs (fig. 11.29). Plesio-

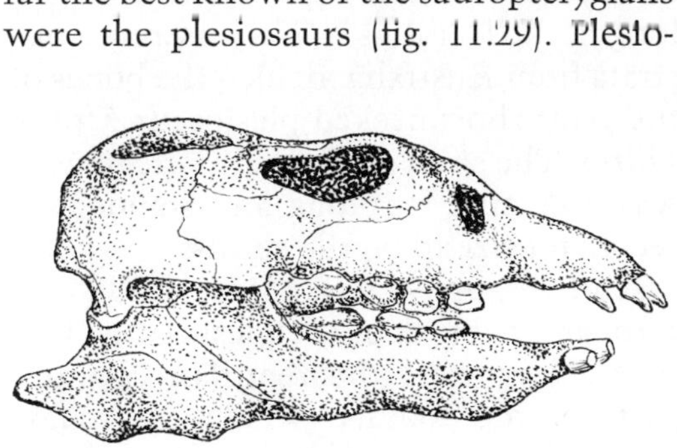

FIG. 11.28 *Skull of* Placodus. *(After A. S. Romer, 1966,* Vertebrate Paleontology *3d ed., University of Chicago Press, p. 125.*

saurs apparently evolved from an early nothosaur species. Their fossils appear for the first time in marine Jurassic strata. Sometimes called swan lizards, they had short, flat bodies; their limbs transformed into broad, many-boned paddles. The neck in some species was very long and snakelike. It terminated in a smallish head armed with long, slender, curved teeth—veritable fish snares. Years ago the paleontologist Dean Buckland aptly described one of these long-necked plesiosaurs as resembling a snake which had been threaded through a turtle. *Elasmosaurus* from the Cretaceous attained an overall length of forty feet. In addition to these long-necked, small-headed plesiosaurs, there was another group with very large heads and short necks. These appear to have been more vigorous divers than long-necked forms. Early Cretaceous strata from Australia yielded the bones of the giant short-necked plesiosaur *Kronosaurus*. The skull alone in this great beast was over nine feet long and stands as a record for size in reptile skulls.

FIG. 11.29 *The plesiosaur,* Elasmosaurus.

Ichthyosaurs were far more fishlike in form and habit than the plesiosaurs. In a sense, they were the reptilian counterparts of the toothed whales of today. They varied in size from small porpoise-like forms to saurian Moby Dicks over thirty feet in length. A fishlike hypocercal tail, dorsal fin, and paddle limbs extended from the streamlined body. Even the vertebrae resembled those of fishes and permitted the necessary undulations needed for swimming. The ichthyosaur head was a suitably pointed entering wedge. A ring of body plates surrounded the large eyes and may have provided protection against pressure during deep dives. Although a few forms were toothless, most had sharp teeth set in long, pointed jaws. They were active predators, with good vision and ample speed for the hunt. Some apparently bore their young alive, for adult specimens have been discovered which show the young partially emergent from the pelvic area of the mother. *Ichthyosaurus* (fig. 11.30) is a representative form which displays well the peculiar downward bend of the tail vertebrae which characterized this group.

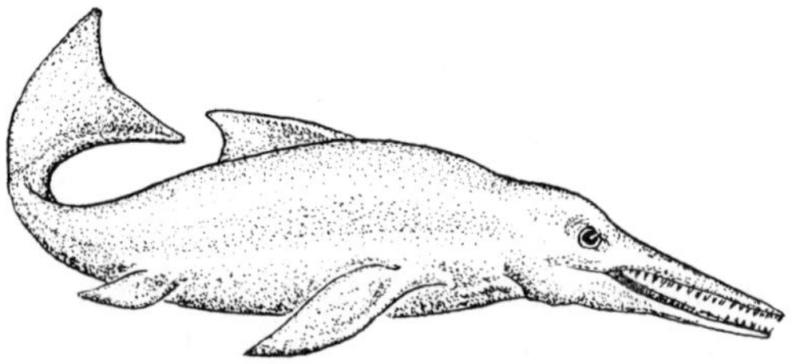

FIG. 11.30 Ichthyosaurus.

By Jurassic time, seagoing crocodiles of the diapsid subclass had appeared. Some grew to lengths of over fifty feet and used their long, slender, tooth-lined jaws for capturing fish and squid. *Geosaurus* (fig. 11.31) and other members of his taxonomic family developed a reverse heterocercal tail similar to that found in the ichthyosaurs. Although the geosaurs were apparently well suited for life in the sea, their history was not a long one,

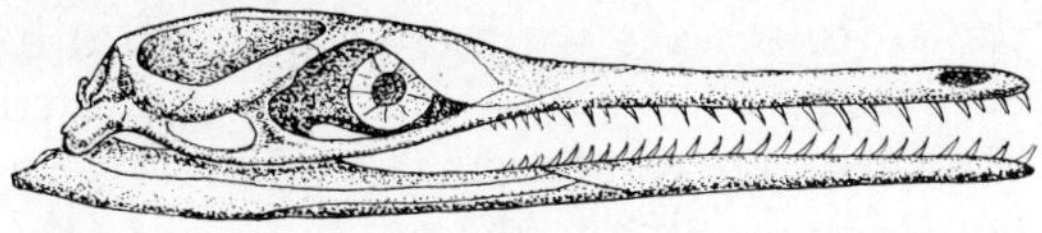

FIG. 11.31 *Skull of* Geosaurus. *(After A. S. Romer, 1966,* Vertebrate Paleontology, *3d ed., University of Chicago Press, p. 141.)*

for only a very few were left at the beginning of the Cretaceous. Even these soon disappeared. It is at least a possibility that their place was taken by those great marine lizards, the mosasaurs.

The mosasaurs were descendents of a group of relatively small aquatic monitor lizards. Like the marine crocodiles, they were diapsids. They were very successful marine reptiles during the latter part of the Cretaceous, and excellent museum specimens have been obtained from the chalk formations of Kansas. The mosasaurs take their name from the Meuse River (Latin, *Mosa*) in Holland where the first specimen was discovered. In 1795, the same fossil was appropriated by the French revolutionary army and sent to Paris. A typical mosasaur was shaped somewhat like a greatly enlarged moray eel to which was added four broad paddles for steering. Propulsion was accomplished by the sculling action of the long vertically flattened tail and undulations of the body. *Tylosaurus* (fig. 11.32) represents the more advanced mosasaurs that grew to lengths of over twenty feet. The most unusual part of the mosasaur skull was the lower jaw. Instead of being fused into a rigid structure and hinged at the rear, the jaw was also hinged in the middle, so as to greatly increase its flexibility and gape. Mosasaurs were primarily fish eaters, but some were adapted to a diet of molluscs. Shells of cephalopods have been found with serially arranged puncture marks that precisely correspond to the dental pattern of their mosasaurian enemies.

FIG. 11.32 *The mosasaur* Tylosaurus.

The euryapsid plesiosaurs, ichthyosaurs, and placodonts, as well as the diapsid marine crocodiles and mosasaurs, shared the Mesozoic seas with a perhaps less spectacular, but nevertheless highly persistent group of anapsids—the sea turtles. Cretaceous turtles followed the trend toward giganticism established by these other groups. *Archelon* (fig. 11.33) attained a length of twelve feet. As an adaptation to their aquatic environment, the shells of marine turtles tended to thin

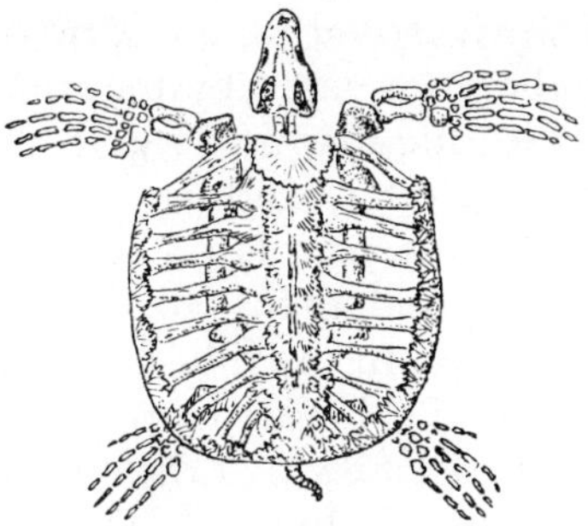

FIG. 11.33 *Skeleton of the Cretaceous sea turtle* Archelon *viewed from above (ventral plastron not shown).*

out and disappear. *Archelon* had only a sketchy framework of a shell, a trait which the living marine leatherback turtles carry even further. Most sea turtles have horny beaks and tough throat linings which assist them in feeding on abrasive plants and invertebrates that would not be edible by many other reptiles. The turtles have been a very persistent race. They have occupied a niche with relatively few direct competitors and are a tolerant group.

ALL THAT REMAIN

As we have seen, there was a great fullness of life during the Cretaceous. It was the time of the dinosaurian heyday. In addition, angiosperms expanded across the surfaces of the continents and stimulated changes in the world's plant eaters. There were also small furry mammals, turtles, snakes, crocodiles, frogs, and lizards. It was for the most part the best of all reptilian worlds. Then, however, came the great biological crash which marked the close of the Cretaceous. All of the dinosaurs disappeared, as did the flying reptiles, great marine reptiles (except sea turtles), molluscs such as ammonites and belemnites, and a host of entire families of other animals. The world henceforth would be a vastly different kind of place, soon to be dominated by mammals, and eventually, the mammals with the oversized brains. Yet, not all reptiles were lost. Turtles, snakes, lizards, crocodilians, and the rhynchocephalian called *Sphenadon* survived. Even in the midst of challenges of the so-called "higher vertebrates," they have persisted down to the present day.

Turtles are chelonians. They first appeared in the Triassic as inhabitants of freshwater bodies. Some then colonized the land, becoming the forerunners of the tortoises. As we have seen, still others returned to the sea and evolved into the great sea turtles. Turtles are considered direct descendents of the cotylosaurs and may have evolved from a form not unlike the Permian *Eunotosaurus* (fig. 11.34). This creature measured only a few inches in length and had its eight ribs expanded so as to form a carapace or dorsal shell. *Proganochelys* (fig. 11.35) of the Triassic had not yet developed a horny beak, nor could this animal pull his neck into his shell. Later turtles developed the retractable head and legs, and the beaklike, toothless jaws. Of course, the protective carapace is the

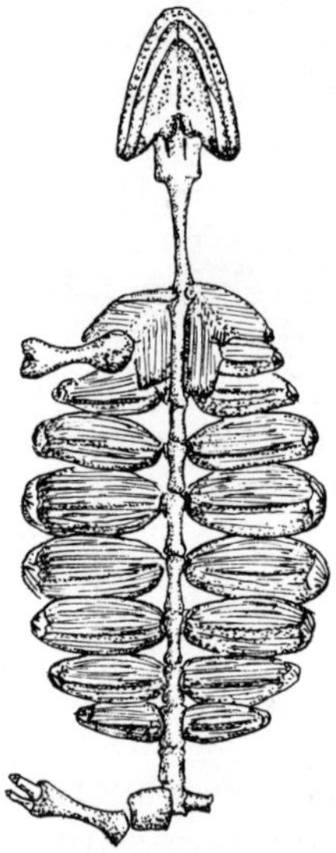

FIG. 11.34 *Partial skeleton of* Eunotosaurus.

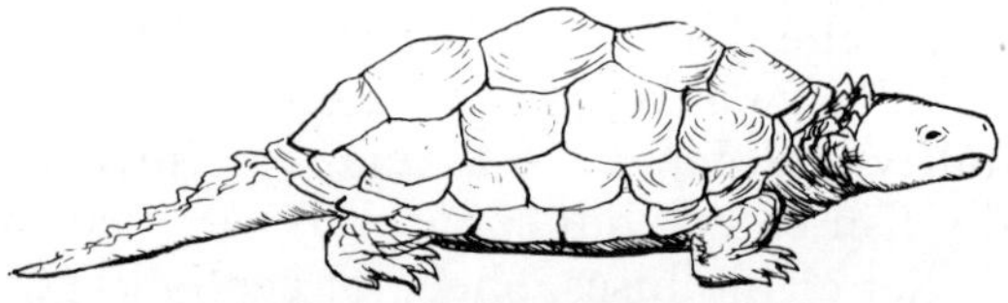

FIG. 11.35 Proganochelys.

most distinctive chelonian trait, and undoubtedly contributed significantly to their long survival.

The armor of a turtle has two layers, an outer layer or dermal plates and an inner layer of bony plates. For added strength, the joints between the plates do not coincide. The upper or dorsal armor comprises the carapace, and the lower shell is the plastron. Carapace and plastron are connected to one another by "bridges" at the sides. Finally, the carapace is fused to the vertebrae along its central dorsal keel.

The more primitive of living turtles are the *Pleurodira,* or "side necked" turtles, that inhabit freshwater bodies in Southern Hemisphere continents. In these turtles, the head is drawn into the shell by tucking the neck in sideways under the front edge of the carapace. Far more numerous in the modern world are the *Cryptodira,* or "concealed neck" turtles and tortoises. In cryptodires, the head is withdrawn straight back under the front edge of the carapace in a vertical sigmoidal flexure. Naturally, the cervical vertebrae are specialized to make this flexure possible. Cryptodires have been numerous since the Cretaceous and have adapted to almost every environment from deserts to seas. Most are small, but the group does include such giants as the twelve-foot *Archelon* of the Cretaceous, and *Colossochelys* of the Pleistocene. Today we marvel at the much smaller, but still impressive, Galapagos tortoises.

Crocodilians are the largest and most bellicose of the surviving reptiles. With a few exceptions, today's crocodilians live in fresh water, where they depend upon stealth more than speed to catch their dinners. Unwary fish swimming alongside what appears to be a floating log are suddenly snapped up. Often crocodiles lie in wait for land animals that come down to the water's edge to drink, and then the huge traplike jaws snap over the victims. Death usually follows by drowning, as the struggling animal is pulled backward and into the water.

Crocodiles got their evolutionary start in very Late Triassic, with the appearance of *Protosuchus* (fig. 11.36) and his kin.

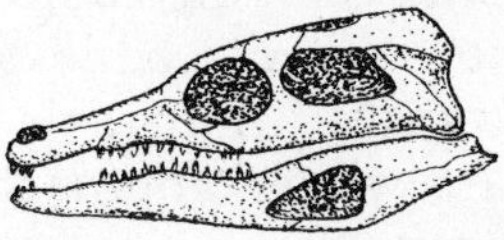

FIG. 11.36 *Skull of* Protosuchus. *(After E. H. Colbert, 1969,* Evolution of the Vertebrates, *Wiley, p. 241.)*

The protosuchians, or "first crocodiles," had rather short skulls, but the structure of the limbs was decidedly crocodilian, and both the back and belly were crocodilelike in their covering of rows of bony plates. By Jurassic time, crocodiles had completely replaced the phytosaurs with which they had convergent similarities. The Jurassic animals were somewhat more primitive than those living today, in that the secondary palate was less well developed. They are collectively termed mesosuchians. The Eusuchia, or modern crocodiles, appeared in the Cretaceous, flourished during the Tertiary, and continue down to the present day. During the Cretaceous, *Phobosuchus,* the "horror crocodile," rivaled some of the dinosaurs in size and may even have made a meal of some of them. It was some forty-five to fifty feet long, and it had a skull measuring over six feet in length.

Alligators, crocodiles, gavials, and caimans are all varieties of modern crocodilians. All are aquatic and relatively large.

Their feet are partly webbed, and they swim by use of these when rapid bursts of speed are not required. For more rapid movement, the laterally flattened tail is used. Unlike the phytosaurs, the nostrils are located high on the anterior end of the skull, and the internal nares are carried far back by the presence of a secondary bony palate. Thus crocodiles can breath under water even when their mouths are open, provided the tip of the snout is emergent. In addition to their hard palate, the crocodile shows such other "advanced" traits as a four-chambered heart which approaches that of birds in structure, richly subdivided lungs, and the presence of a diaphragm (not homologous to the diaphragm in mammals). In spite of these progressive characteristics, however, they clearly retain evidence of their origins from Mesozoic bipedal reptiles in the larger size of their hindlimbs relative to the forelimbs. Of all the surviving reptiles, crocodiles are most closely related to the dinosaurs.

Snakes and lizards are placed together in the Order Squamata. Without doubt, they are the most numerous and varied of living reptiles. Their origins can be traced back to the Triassic, but fossils are rare except for the giant marine lizards, the mosasaurs. Snakes are basically "legless lizards." The evidence for this is apparent in the living pythons, which retain vestiges of hind limbs and pelvic bones. Snakes differ further from lizards, in the extreme structural mobility of the skull. The upper jaw and part of the face can move forward and upward on the cranium to produce an extraordinary gape. The sides of the lower jaw are not sutured, but merely joined by an elastic ligament. This highly flexible arrangement permits snakes to swallow prey much larger than themselves. However, the fragile and loosely joined skull parts do not lend themselves readily to preservation. Nevertheless, both fossil snakes and lizards are known from beds as old as Cretaceous. They evolved into serpents as formidable as the sixty-foot constrictor from the Eocene of Egypt, and a South American poisonous snake with two and a half inch fangs.

The snakes and lizards were offshoots of a small group of lizardlike diapsid reptiles termed Eosuchia. Yet another offshoot of the eosuchians goes by the rather ponderous name Rhynchocephalia. These were mostly lizardlike reptiles that persisted from Triassic to the present. Today, a single endangered species remains; namely, "tuatara" or, more formally, *Sphenodon punctatum* (fig. 11.37). This "living fossil" is rigidly protected and survives on islands off the coast of New Zealand. Tuatara resembles a rather stocky lizard. It is about two feet in length with small spines extending down the length of its back. The teeth are fused to the jaws, rather than set in sockets, and the front of the skull has a short overhanging beak. A primitive feature of interest is the tuatara's pineal eye

FIG. 11.37 *The tuatara* Sphenodon.

—a structure with lens and retina which lies just below the skin on the top of this reptile's head.

The tuatara is an appropriate animal with which to close our study of reptiles. We leave these survivors of the great age of dinosaurs and look forward to a world of warm-blooded creatures with larger brains and with hair or feathers.

SUMMARY

Modern reptiles, although numerous and varied, are in a relatively minor position when compared to the mammals, fishes, and birds. However, during the Mesozoic Era, they were the dominant large land animals of the planet. Unlike the amphibians, reptiles are completely independent of water bodies throughout their life cycles. They produce large eggs that can be laid on land. The egg is covered with a leathery or calcareous "shell" which protects the developing embryo from harm. Food for the embryo is stored in the yolk of the egg, and extraembryonic membranes provide for excretion and respiration. Copulatory organs are developed so that fertilization is generally internal. The reptilian body is protected by thick cornified skin, upon which closely fitted scales are superimposed. In general, reptiles possess more highly developed respiratory and circulatory systems than do amphibians.

The first reptiles evolved in the Carboniferous and began a diversification that produced most of the orders that dominated vertebrate life during the Mesozoic. The early forms, termed cotylosaurs, were in many ways transitional between amphibians and reptiles.

A useful, if somewhat arbitrary scheme for classifying reptiles is based on the presence and locations of openings in the skull to accommodate bulging temporal muscles. Reptiles lacking such openings are referred to as Anapsida. Modern turtles and the ancient cotylosaurs are representative anapsids. As implied by their name, the Diapsida have two temporal openings. A single opening high on the skull identifies the Euryapsida, while Synapsida have a single opening lower on the side of the skull.

The synapsid reptiles have a distinct claim to fame. They are the group that gave rise to the mammals. They include the pelycosaurs ("sail-back reptiles") as well as the therapsids (often designated as mammallike reptiles"). The therapsids possessed several anatomical features suggesting reptilian approaches to the mammalian status, including a degree of tooth differentiation, emphasis of the dentary bone in the lower jaw, closure of the pineal opening, and more vertical alignment of the limbs beneath the body.

The Diapsida include not only living lizards, snakes, and crocodiles, but also the extinct "dinosaurs," mosasaurs, and pterosaurs. The evolutionary pathways that led to the dinosaurs began in Permian time with a primitive group of diapsids called lepidosaurs. Possibly from lepidosaurian stock, the more progressive archosaurian lineages evolved. Thecodonts were a group of Triassic archosaurians from which the dinosaurs arose. According to the orientation of pelvic bones, dinosaurs belong to two major orders; namely, the Saurischia which include the well-known bipedal carnivores and the gigantic swamp-dwelling herbivores, and the Ornithischia in which are placed duckbilled, plated, armored, and horned dinosaurs.

Reptiles with a single, dorsally located

temporal opening—the Euryapsida—were mostly modified for life in the sea. The group includes the sauropterygians (plesiosaurs), placodonts (mollusc-eating reptiles), and the fishlike ichthyosaurs. These successful marine reptiles, along with most of their reptilian relatives ashore, became extinct about seventy milion years ago. The survivors of the great "Age of Reptiles" are the snakes and lizards, the crocodilians, turtles, and the bead-headed "lizard" of New Zealand—the tuatara.

SUPPLEMENTARY READINGS AND REFERENCES

Colbert, E. H. 1965. *The age of reptiles.* New York: Norton, p. 211.

———. 1968. *Men and dinosaurs.* New York: Dutton.

De Camp, L. S. and De Camp, C. C. 1968. *The day of the dinosaurs.* New York: Modern Library Editions Publishing Company.

Jensen, G. L. 1964. Riddles of the terrible lizards. *American Scientist* 52: 227–246.

Kurten, Björn. 1968. *The age of the dinosaurs.* New York: McGraw-Hill.

Romer, A. S. 1968. The origin of the amniotic egg. *Scientific Monthly* 85: 57–63.

Stahl, Barbara J. 1974. Vertebrate history: problems in evolution. New York: McGraw-Hill.

12 Into The Sky

Next to water as a moulding environment comes the air, for the creature which inhabits either is surrounded on all sides by a homogeneous medium, so that it becomes uniformly and beautifully modified to offer the least possible resistance to the attainment of speed.

—Richard Swann Lull, *Organic Evolution*

DESIGN FOR FLIGHT

The final classes of vertebrates to evolve differ fundamentally from all earlier classes. Fishes, amphibians, and reptiles are ectothermic. Their body temperatures vary more or less as the temperature of the surrounding air or water. Birds and mammals are endothermic, or "warm-blooded."

Paleontologists have long been aware of the structural similarities between birds and reptiles. Instructors in biology classes are fond of referring to birds as being "little more than feathered lizards." The expression, however, does little justice to the marvelous attainments of birds, not the least of which are warm-bloodedness and superior powers of flight. Both of these attributes are related to the transformation of feathers from what once were reptilian scales. In addition to their obvious importance in flight, feathers are an ideal insulation, and help to maintain the high avian body temperature. For their weight, feathers are sturdier than any flight surface designed by man. They are also flexible, bending upward with each downstroke of the wing thereby producing the equivalent of pitch in an airplane propeller and providing bouyancy and forward propulsion. Birds have been fashioned by nature for high power and low weight. Feathers, hollow bones, efficient wings powered by sturdy breast muscles, a large powerful heart, and a remarkable respiratory system all combine to produce an aerial acrobat that never fails to inspire wonder in the eyes of inventive man. Although the skeleton commonly weighs less than 5 percent of body weight, it is nevertheless both strong and elastic. Strength is added by the fusion of some of the bones ordinarily separated in other vertebrates. Diagonal braces resembling the struts used in airplane wing design are found in the larger bones of some birds.

The high rate of metabolism in birds is well known. Even the embryo must be incubated, and parental care is needed after hatching. Body temperature in adult birds is higher than in any other animal, often reaching 113°F. Driven by the pulses of the strong four-chambered heart, the blood circulates rapidly and makes a purifying side trip through the lungs before moving through the body again. High energy food is utilized and quickly processed. To help cool the body and add to the efficiency of respiration, birds have an auxiliary system of five or more air sacs that ramify through the body. Indeed, birds maintain their intense level of activity at the expense of a long life. Small songbirds rarely live beyond two years. Perhaps this is a small price to pay for the power to soar above earthbound creatures. It is an enviable gift. Yet man need not be saddened, for the miracle of evolution provided him with hands rather than wings. Hands have taken man beyond the realm of eagles into the vastness of space.

THE LIZARD BIRD AND THOSE THAT FOLLOWED

The fossil record for birds is not good. Small animals with thin bones are not readily fossilized. The record, inadequate as it is, seems to indicate that birds remained fairly limited in numbers and variety until the Cenozoic. Perhaps they were unable to achieve their full adaptive potential until after the reptiles had begun to dwindle.

It appears certain that birds evolved from Triassic thecodonts; however, the precise ancestor from among the thecodonts has not been found. As we have previously noted, the thecodonts were already birdlike in their hollow bone structure and bipedal stance. The earliest bird fossil, *Archaeopteryx* (fig. 12.1), is a perfect link between these reptiles and modern birds. The earliest evidence of the existence of *Archaeopteryx* was the 1860 discovery of a single feather clearly imprinted in a slab of lithographic limestone from the quarries at Solnhofen. A year later, the formation yielded a beautifully preserved specimen that was complete except for the fact that the head was missing. Another six years passed before a skeleton with head intact was discovered. A very few fragmentary and poorly preserved fossils have been found since that time. Together, they permit us to clearly reconstruct the appearance of this link between the reptiles and the birds.

FIG. 12.1 Archaeopteryx, *the first bird.*

Archaeopteryx was a creature about the size of a crow. Except for feathers, most of its features were reptilian. For example, the jaws bore reptilian teeth, the tail was long and lizardlike (with a row of feathers on either side), and the vertebrae lacked the distinctive modifications characteristic of birds generally. Unlike the wings of modern birds in which the fingers tend to coalesce for greater strength, the wings of *Archaeopteryx* had handlike, claw-bearing, free digits to assist the opposable first finger in climbing and grasping. The presence of these projecting fingers supports the interpretation that these creatures were forest dwellers that glided from one tree branch to another and were then able to scramble quickly up or down with the help of their clawed hands. This is not the manner of a vigorous aviator, and it seems *Archaeopteryx* was still in the experimental stage of flight design. The animal did possess the bone popularly known as the "wishbone," which is characteristic of birds. However, the sternum lacked a keel, suggesting that massive breast muscles needed for long distance flight were lacking.

As paleontologists have not yet discovered the Triassic ancestor of the "lizard bird," we can only surmise the sequence of events that led to feathered flight. It all began, of course, with the evolution of feathers from scales. The original purpose of the feathers may have been more for insulation than flight. Gradually the longer feathers became concentrated along the forelimbs, which were then utilized as gliding surfaces. One can at least speculate that the lift provided by the feathers would be quite an advantage to a lizard that leaped from branch to branch. With the passage of time and many generations, the tail was reduced, a strongly keeled breastbone developed to anchor the sturdy wing muscles, and the beak replaced the toothed jaws.

The Cretaceous of Europe and North America provides the next glimpse of bird evolution. The best-known genus from the Cretaceous is *Hesperornis* (fig. 12.2). Called "the bird of the west" because its remains are found in the Niobrara Chalk of Kansas, *Hesperornis* was a splendid, aquatic, toothed bird measuring over four feet in length. It fed on fish captured while swiming and diving. *Hesperornis* had completely lost the ability to fly. This is evident from the slender solitary humerus in the wing and the greatly reduced shoulder girdle. Webbed feet on sturdy rear limbs provided the propulsion. Sea birds that were contemporaries of *Hesperornis* had essentially modern flying powers. They resembled the modern gull or tern, except that the jaws still bore teeth. Clearly, the record for aquatic birds is superior to that of terrestrial birds. The basins of the sea have always provided a more secure graveyard.

FIG. 12.2 Hesperornis, *a flightless Cretaceous marine bird.*

The Cenozoic fossil record for birds is also extremely sparse. Fragmentary evidence suggests that birds were toothless and essentially like those of today, and included a variety of woodland, flightless, and predatory birds. Large flightless birds, like the extinct moa and *Aepyornis* and the living ostrich and rhea, have been important Cenozoic animals. Some of these seem to have developed initially in localities in which potentially dangerous carnivores were lacking, such as South America, Australia, and New Zealand. The Cenozoic also seems to have been the time of development of great speed in flight, as well as the miraculous and still not completely understood habit of migration. Perhaps too, song may have first occurred among these Cenozoic birds.

Birds have always been a compact and highly specialized group. Their range of size and adaptations are not as great as in the mammals and reptiles. They are marvels of adaptive perfection, specialists rather than generalists. In their special way, they have helped to create a more gracious and comfortable habitat for mankind.

SUMMARY

Birds are all animals with feathers. Although we are accustomed to thinking of feathers only as necessary structures for flight, they also play an important role in insulating the bird against loss of body heat.

In addition to the unique covering of feathers, birds have hollow bones, a powerful four-chambered heart, strong wings moved by sturdy breast muscles, a remarkable respiratory system, and a high rate of metabolism. It is the combination of all of these features that makes efficient flight possible in vertebrates. Birds achieved a fairly wide adaptive radiation. Some have even lost the ability to fly, although they still retain wings which no longer function for flight.

All that is required in order to distinguish a fossil bird is the impression of a feather. Feathers are believed to have evolved from scales among the reptilian ancestors of the birds. The earliest known bird, *Archaeopteryx,* was found in the Jurassic Solnhofen limestones. Except for the feathers, the anatomical features of this creature were clearly reptilian. During the Cretaceous, flightless aquatic birds like *Hesperornis* fished the inland seas. Fossil remains of Cenozoic birds are sparse, but fragmentary fossils do indicate that birds were completely modernized from the beginning of this final era in earth history.

SUPPLEMENTARY READINGS AND REFERENCES

Darling, Lois and Darling, Louis. 1962. *Birds.* Boston: Houghton Mifflin, 261 pp.

Heilmann, G. 1927. *The origin of birds.* New York: D. Appleton, 210 pp.

Lanyon, W. E. 1963. *Biology of birds.* New York: American Museum Science Books, The Natural History Press, 186 pp.

Storer, John H. 1952. Bird aerodynamics. *Scientific American* 186 (4): 24–29.

Swinton, William E. 1965. *Fossil birds.* London: British Museum of Natural History, 65 pp.

Welty, Carl. 1955. How birds fly. *Scientific American* 192 (3): 88–96.

13 Mammals Take The Field

Increasing development of grasping feet
in primates excluded for them any immediate
opportunity to match the fleetness of the
deer, but it led to opportunities whose
seizing eventuated in the finger on a trigger,
against which the deer's fleetness is futile.
—G. G. Simpson, *The Meaning of Evolution*

AN UNPRETENTIOUS DEBUT

Mammalian history begins with the appearance of the mammallike reptiles in the final days of the Paleozoic Era. Earlier in that era, climates had been warm, humid, and generally friendly to life. As conditions became more severe in the Permian and Triassic, swamp-dwelling vertebrates diminished, but the expanding upland areas favored the adaptations that led to the radiation of the therapsids. From the Karoo beds of Africa, we find the bones of paramammals that had almost crossed the reptile-mammal boundary. That boundary is in many ways transitional and arbitrarily defined. It is usually drawn where there is evidence that the reptilian quadrate-articular jaw articulation has been modified to the mammalian squamosal-dentary type of hinge. In the process, the bones of the older articulation become mammalian ear ossicles.

When compared to the quantity of Mesozoic reptile remains already collected, fossils of the earliest mammals are indeed rare. A few localities contain pockets of sediment which, on sieving, yield tiny bones and teeth. One such locality is a Late Triassic deposit in South Wales where the remains of mouse-sized *Morganucodon* (fig. 13.7), the earliest known mammal, was discovered. Before examining the traits of these primitive mammals, however, it is necessary to consider the more general attributes of the class of vertebrates to which we ourselves belong.

Mammals are familiarly recognized by the possession of fur or hair and, as implied by their name, by mammary glands used in suckling their young. The first of these traits is associated with warm-bloodedness and the second with a measure of postnatal care. Neither characteristic, however, is very useful to the paleontologist who must work with harder materials. There are several characteristics in the mammalian skeleton that are especially diagnostic. One of these is the presence of a solitary bone, the dentary, in the lower jaw. As noted previously, the malleus and incus are also uniquely mammalian ossicles, but are so minute they are rarely found at fossil sites. The middle ear in mammals is surrounded by the tympanic bone which may be ringlike, but more often is expanded into a hemispherical shell called the bulla. The primitive mammalian skull is fundamentally synapsid, but with a greatly expanded braincase. A double occipital condyle forms the articulation between the skull and the first cervical vertebra. In mammals, a secondary palate separates the nasal passage from the mouth, and there is a single nasal opening in the front of the skull. This orifice contains turbinal bones whose function is to support the folds of mucous membranes which warm air as it is inhaled. There is ordinarily a functional differentiation of kinds of teeth in mammals. In man's dentition, this differentiation consists of incisors, canines, molars, and premolars. Other differences in mammalian skeletal anatomy are found below or posterior to the skull. The cervical vertebrae appear to be ribless because the ribs are greatly reduced and amalgamated into an integral part of each vertebra. The more posterior lumbar region becomes ribless and forms a contrast to the great rib basket of the thorax. Characteristically, the number of vertebrae in each region of the skeleton are fewer and less variable in shape than in the reptiles. As an example, in all but four genera of living mammals, the number of cervical vertebrae is seven, regardless of the length of the neck. The breastbone (sternum) which was relatively unimportant in the reptiles, becomes an important base of attachment for the longer dorsal ribs. In mammals one also finds a strong spine extending down the middle of the shoulder blades. This feature provided a place for the attachment of muscles from the upper forelimb and is related to the mammalian trait of positioning the legs more directly beneath the body. Bones of the pelvis (ilium, ischium, pubis) are fused into a single sturdy structure, and there is a reduction in the number of toe bones. The first digit or thumb possesses two phalanges, and each of the other four digits only three phalanges each.

One can only surmise that the earliest true mammals had essentially the same soft-part anatomy as do most living mammals. We infer the early existence of the mammalian four-chambered heart as being necessary for a warm-blooded animal with a high level of activity. This vital organ actually has, in mammals, two pumps which operate in unison. The two chambers on the right serve as one pump sending blood from the body circulation to the lungs. The other two chambers pump blood from the lungs back into the body network.

The brains of reptiles are, as we have seen, tiny masses tucked away in an inconspicuous enclosure at the rear of the skull. By comparison, the brain of mammals is enormous. Eyes, ears, and especially noses are important sense organs. Indeed the cerebral hemispheres, which were originally small structures dedicated to an olfactory function, are greatly enlarged and from them arise the higher brain centers and improved mental endowments of the mammals.

CATEGORIES OF MAMMALS

According to their level of organization and method of reproduction, living mammals can be grouped into three categories:

the placentals, marsupials, and monotremes. Placental mammals are the most advanced and, today, the most numerous of the three. In placentals, the embryo is internally attached to the mother by a structure known as the placenta, from which the name of the class is taken. The placenta is formed partly from the allantoic membrane and partly from the uterine wall of the mother. The fetus receives food and oxygen through the placenta, and waste products are carried away through the mother's blood. On birth, the mother feeds milk to the young through teats located on the ventral surface.

Mammals called marsupials have teats or mammae, but no placenta—or only a very ephemeral one. The young are born prematurely, then move into a fold of the body wall in which are located the nipples of the mammae. The premature marsupial becomes firmly attached to a nipple and milk is pumped into his body. He remains in the pouch until sufficiently developed to venture out into the world. Today the marsupials are represented by such animals as the kangaroos, wallabies, wombats, phalangers, bandicoots, and koalas of Australia; and opossums of the new world. They are a dwindling group.

There are several skeletal traits which are useful to the paleontologist in distinguishing marsupials from placental mammals. The posterior angle of the lower jaw bends inward beneath its articulation with the skull in a manner which has come to be known as the marsupialian "inflected angle" (fig. 13.1). In addition, the braincase is, on the average, smaller than in placental mammals of equivalent size. The molar teeth have an archaic triangular pattern of cusps, and whereas

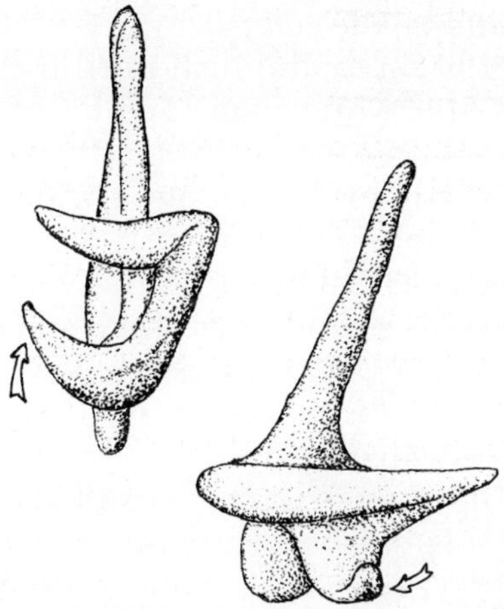

FIG. 13.1 *Marsupial and placental lower jaws viewed from rear. Upper left is marsupial (opossum), lower right is placental (mink). Arrows point to angular processes. (Adapted from J. M. Weller, 1969,* The Course of Evolution, *McGraw-Hill, p. 491.)*

placentals have three premolars and four molars. Epipubic or "marsupial bones" for support of the pouch are found in the pelvic region. Many marsupials show a peculiar specialization of the foot, in that the second and third toes are modified into tightly appressed slender digits—called syndactylous digits—which are used as a comb for grooming the fur (fig. 13.2).

It is well known that Australia and South America are the continental abodes of large numbers of marsupials.

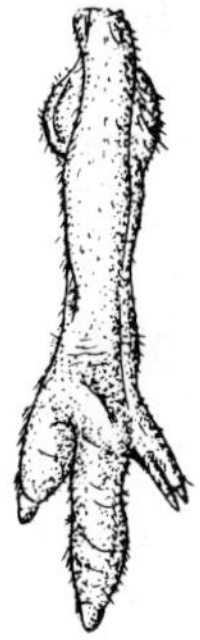

FIG. 13.2 *Hind foot of kangaroo showing syndactylous digits.*

In Australia, the great array of marsupials did not begin to diminish until the coming of the white man. Fossils of Australian marsupials older than Pleistocene are virtually unknown, but a wide range of adaptations probably existed before the Ice Age as is indicated by the variety of the survivors. Today we find marsupial carnivores (tasmanian devil, wolf, and cat), moles, insectivores, and herbivores like the wallabie and kangaroo. Apparently, the Australian marsupials were more isolated and secure on their island continent than the South American forms. The latter were confronted with adverse competition once the connection between North and South America was established late in the Pliocene.

The South American marsupials sprang from opossumlike ancestors. They are represented in the fossil record primarily by flesh-eating animals. In Miocene time, the wolflike *Boryhyaena* (fig. 13.3)

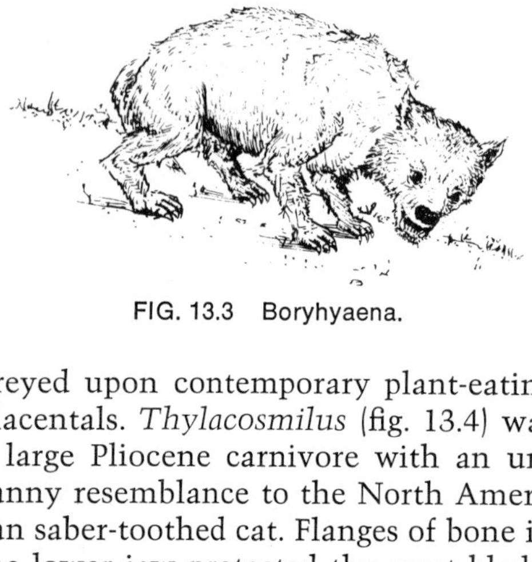

FIG. 13.3 Boryhyaena.

preyed upon contemporary plant-eating placentals. *Thylacosmilus* (fig. 13.4) was a large Pliocene carnivore with an uncanny resemblance to the North American saber-toothed cat. Flanges of bone in the lower jaw protected the great bladelike canines from injury when the jaws were closed. Yet another marsupial, the caenolestoid, adapted to the insectivore niche.

FIG. 13.4 Thylacosmilus, *the marsupial sabertooth.*

Not all of the South American mammals were marsupials. A small fauna of primitive plant-eating placentals had found their way into the continent prior to its separation from North America at the beginning of the Tertiary. As we shall see shortly, these plant eaters (along with some of the marsupial carnivores) evolved into a fauna that remarkably paralleled the radiation of mammals elsewhere in the world. The two new world continents hosted animals that superficially duplicated the general morphology and habits of horse, rhinoceros, camel, dog, cat, shrew, and many others. Unfortunately, the South American partners in the duplication did not fare well when confronted with migrants from the north.

There are a few species of mammals that belong neither to the placentals nor the marsupials. These relics of long ago laid eggs in the reptilian manner. They are called monotremes. However, unlike reptiles, monotremes possess primitive mammae that have been modified from sweat glands and provide the newly hatched offspring with nourishment, if only for a relatively brief period. The fossil record for these most primitive of living mammals begins in the Pleistocene, although it seems certain they were present much earlier. Only three monotremes still live today; namely, the highly specialized, stream-dwelling duck-billed platypus (fig. 13.5) of Australia and Tasmania, and two species of spiny ant-eaters of Australia and of New Guinea (fig. 13.6). These animals retain many

FIG. 13.5 *Platypus.*

reptilian features both in soft organs and skeleton. However, they are decidedly mammalian in the characteristics of the lower jaw and ear, the possession of seven cervical vertebrae, a diaphragm, the single left systemic arch of the heart, hair, and organs for suckling. Although surviving forms are highly modified for special modes of life, the monotremes represent an evolutionary level only

FIG. 13.6 Echidna.

slightly above that attained by the mammallike ictidosaurian reptiles of the Late Triassic. However, monotremes are not considered ancestral to the other mammals. Rather they are an entirely separate line of evolution that arose independently from some of the mammallike reptiles about two hundred twenty-five million years ago.

THE MAMMALIAN VANGUARD

Mammals presumably came into existence during the Late Triassic, but knowledge of their Mesozoic history is severely limited by the sparse and fragmentary nature of the early fossil record. From time to time teeth, jaws, or parts of skulls are found, but not complete skeletons. These scrappy remains permit recognition of several groups of primitive Mesozoic mammals. Each is recognized primarily on the basis of differences in tooth morphology. They are named the Morganucodonta, Docodonta, Triconodonta, Symmetrodonta, Pantotheria, and Multituberculata.

The earliest of the mammals were the morganucodonts. They were tiny creatures with a skull that was very close to therapsid in appearance, and a lower jaw that still retained vestiges of reptilian structures. However, the primary jaw articulation was at the squamosal to dentary point, and the lower jaw was functionally a single bone. No doubt, *Morganucodon* (fig. 13.7) of the Late Triassic was close to the reptile-to-mammal transition.

Also close to the boundary between the reptiles and mammals were the doco-

FIG. 13.7 Morganucondon.

donts. They are known from incomplete remains found in North American Upper Jurassic deposits. The crown surfaces of the molars in docodonts are irregularly quadrate in shape and are used for identification of members of this group.

The triconodonts can be recognized by cheek teeth which have three cusps in a longitudinal row (fig. 13.8). Their dental

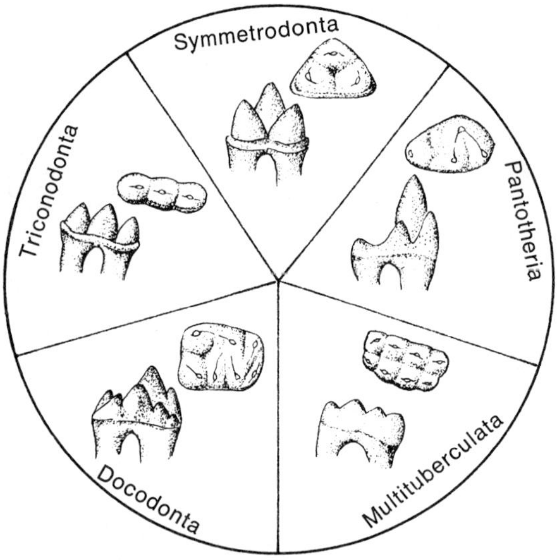

FIG. 13.8 *Molar teeth of Mesozoic mammals. Side views of lower molars as seen from inside and top views of oral surfaces.*

formula is 4–1–4–5, meaning that in each dentary bone there were four incisors, one canine, four premolars, and five molars. These animals were about the size of a cat and probably preyed on smaller reptiles and mammals. They survived into the Early Cretaceous.

The multituberculates range in age from Jurassic to Eocene. Their skulls bear a superficial resemblance to those of rabbits. There are broad chisellike incisors, a gap or diastema between the incisors and cheek teeth, and broad attachment bases for powerful jaw muscles of the kind found in gnawing animals. Multituberculates appear to have been the first entirely herbivorous mammals. They were able to carry their rodentlike adaptations through nearly ninety million years of vertebrate history. *Taeniolabis* (fig. 13.9) is a rather well-known

FIG. 13.9 *Head and lower jaw of the rodentlike multituberculate* Taeniolabis, *jaw about 5 inches long.*

multituberculate that was about the size of a modern beaver.

The symmetrodonts (fig. 13.8) and pantotheres (fig. 13.10) are two rather similar Jurassic orders. Both are represented by small animals that have cheek teeth with a triangular surface. In the symmetrodonts, the cusps are symmetrically arranged at the apex of each corner.

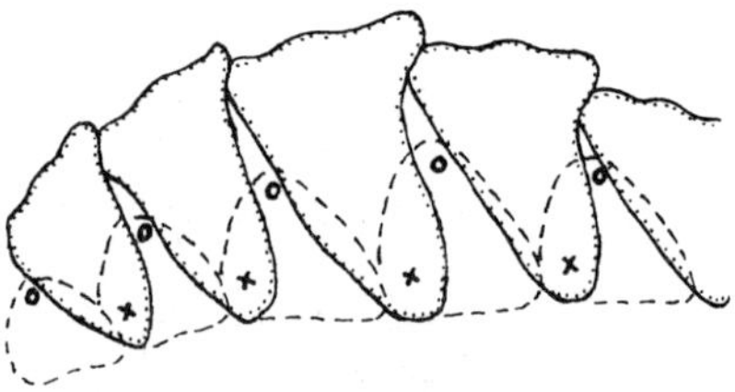

FIG. 13.10 *Plan view of posterior pantothere molars showing reverse orientation of molar triangles, cusps of lower teeth (small circles) fit into spaces between uppers, and inner cusps of uppers (small crosses) meet a heel at the rear of the lowers. Upper teeth solid lined, lower dashed.*

In the pantotheres, this symmetry is less pronounced in that there are extra smaller cusps on the inner border and a "heel" or talonid on the back of each lower molar. There have been many studies based on dental characteristics that attempt to link the pantotheres to the marsupial and placental mammals. The monotremes are, as we have suggested, a separate lineage that may have seen its inception among the docodonts.

THE MAMMALIAN EXPANSION

Although the greatest radiation of mammals was to occur in the Cenozoic, the Mesozoic had been their time for experimentation. For approximately one hundred million years they met the competition of their reptilian contemporaries by developing increasingly more efficient nervous and reproductive systems, greater speed and agility, and a more reliable system of body temperature control. In the following Cenozoic Era, an entire world lay before them. They vigorously expanded into it, embracing not only the habits of their reptilian predecessors, but finding new pathways of adaptation. By Paleocene time, eighteen orders had appeared. Nine of these had already become extinct by Eocene time, and the other nine orders were supplanted by more progressive groups. The phenomenal advance of essentially modern placentals had begun. Altogether it has produced over twenty-eight orders of placental mammals. The list is far too lengthy to treat each great order in detail, but a brief review will hopefully provide insight into the range of structural variation and adaptive inventiveness among these most modern of vertebrates.

Insectivora

The first placentals seem to have made their appearance during the Cretaceous. They were unspecialized insectivores. From this base, the great array of placental mammals began. Insectivores are represented today by small, often nocturnal forms like the shrew and mole. Worms, insects, and slugs are the usual food for these animals. Although some are specialized for particular modes of life, they retain many of the primitive characteristics of their Mesozoic ancestors. They have, for example, unspecialized feet with five digits, and the skull has features characteristic of marsupials, yet it lacks the inflected angle. The dentition is primitive with the basic 3–1–4–3 formula. However, some insectivores show variations in these characteristics that reflect their special ways of making a living. The shrew (fig. 13.11), for example, has developed a flexible snout and reversed the usual trend in evolution by becoming smaller instead of larger. Today they are the smallest of mammals, weighing less than an ounce and measuring under an

FIG. 13.11 *Tree shrew. (From A. L. McAlester, 1968,* The History of Life, *Prentice-Hall, p. 126.)*

inch in length. Shrews are incessantly active and extremely nervous creatures. They have been known to die of fright during thunderstorms. Owing to their small size, they have large surface area relative to body weight. This results in rapid loss of body heat. They must, therefore, eat voraciously, consuming the equivalent of their body weight each day. Small size and secretive habits were the assets that insured the survival of the insectivores for nearly seventy million years. *Zalambdalestes* (fig. 13.12)

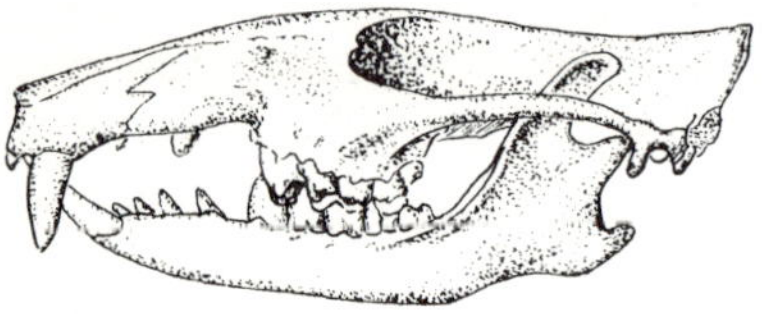

FIG. 13.12 Zalambdalestes.

from the Cretaceous of Mongolia may be considered a representative fossil insectivore. Living insectivores other than shrews and moles, include the tenrecs of Madagascar, old world hedgehogs, and elephant shrews. These are not animals one sees every day, and one must remind oneself of the importance of the insectivores. They are the creatures from which our own order of primates ascended.

Aerial Mammals

Closely related to the insectivores are the bats and so-called flying lemurs. As a matter of fact, these two groups can be thought of as insectivores that learned to fly. The name "flying lemur" is inappropriately applied to the colugo, the only living species of the Order Dermoptera. In truth, the animal does not actually fly, nor is it a lemur. It glides from branch to branch on broad folds of skin extending between the legs and tail. Colugos live only in the East Indies. A few fossils from the Paleocene and Eocene of North America have been found that may represent early relatives of the colugo.

Bats probably evolved from arboreal insectivores. Their aerial adaptation may have begun with a flap of skin (ptagium) between the legs and body much like that seen in the colugo. We do not know that this was positively so, for the oldest fossil bats are Eocene in age and are already distinctly batlike. Bats are the only mammals to have achieved true flight. Their wings are supported by both front and rear limbs, but primarily by the elongated second, third, fourth, and fifth fingers (fig. 13.13). Strong pectoral

FIG. 13.13 *Fruit bat.*

muscles are attached to a small keel on the sternum. Most bats are night creatures and have adjusted to the problem of difficulty in seeing at night by developing a highly efficient sonar system. As they fly about in the darkness, they emit shrill cries that are inaudible to the human ears. These sounds are reflected from nearby objects and picked up by the ears. Modern bats are mostly insectivorous. However, there are also fruit-eating bats which have faces like foxes and therefore, have been labeled "flying foxes."

One also finds living fish-eating, flower-eating, and blood-lapping bats. The latter are the vampire bats which nick the skin of sleeping animals and lap up the blood that flows from the small wound.

Taeniodonts and Tillodonts

The early descendents of the insectivores included two orders that were short-lived. These were the taeniodonts and tillodonts. The former, exemplified by *Stylinodon* of the Early Tertiary, were rather large animals for their time. Their skull was about a foot long with powerful jaws in which were located rodentlike incisors. The legs were short and stocky and terminated in clawed feet. The taeniodonts never became very numerous. They were plant eaters that were soon replaced by the later, more progressive herbivores of the Tertiary. Rodentlike incisors are also found in the tillodonts. *Tillotherium* was about the size of a bear and may have resembled that animal in its ambling gait. Like the taeniodonts, tillodonts were experiments in evolution that suffered early extinction.

Edentates and Other Eaters of Ants and Termites

The mammals known as edentates include such unusual living animals as armadillos, tree sloths, and South American anteaters. Edentate means "without teeth." The name is not entirely appropriate, for some edentates do have simple teeth on which there is no enamel outer covering. Although the various edentates appear quite different outwardly, they do have skeletal similarities. There are extra attachments between vertebrae that are quite unlike any found in other mammals. Edentates do not always have the standard number of seven cervical vertebrae. The geographic origin of the edentates is uncertain. A few rare Early Tertiary fossils found in North America have some armadillo traits, but these so-called palaeanodonts may be a sterile side branch and not directly ancestral to later edentates. A more continuous record of edentates is found among the more numerous South American fossils. These seem to have arisen from armadillolike stock. Armadillos are burrowing creatures that feed on insects, worms, snakes, and carrion, and whose most distinctive characteristics are protective bony plates in the skin which are overlain by horny scales. Fossil species are known from the Paleocene to the Recent.

Closely related to the armadillos are bizarre Late Eocene to Pleistocene edentates called glyptodonts. They were armored forms also, but the armor consisted of a single ornamented carapace. Thus these animals could not roll up into a ball like the armadillos. *Glyptodon* (fig. 13.14) was a veritable walking fortress with a spike covered knob at the end of its armored tail for bludgeoning pesky carnivores.

Sloths are perhaps the most peculiar of the edentates. Today they are found living upside down in tropical forests where they feed on leaves while hanging from

FIG. 13.14 *Glyptodont.*

their large curved claws. The tree sloths move very slowly with a hand over hand motion, feeding on leaves as they inch along. To even further increase the strangeness of their appearance, their fur is frequently colored green with a coating of algae. The tree sloths are living members of a lineage that included giant ground sloths such as the Pleistocene *Megatherium* (fig. 13.15). These animals walked on the sides of their hind feet and knuckles of their front feet in what seems a most ungainly manner.

FIG. 13.15 *Ground sloth.*

The anteaters (fig. 13.16) are another category of interesting edentates. Also South American animals, anteaters have a long snout with barely moveable jaws, and a long, sticky tongue with which they collect the ants and termites on which they feed. Anteaters need no teeth, and so they have none. They have three large claws and one small one on each forefoot, and five claws on the rear feet. The claws are so large that these edentates must walk on the knuckles of their forefeet and—in the fashion of the sloths—on the sides of their hind feet. We know very little about the evolution of anteaters because of the lack of fossils older than Late Tertiary. Like the sloths, they may have developed from early armadillos before the appearance of dermal armor.

FIG. 13.16 *South American anteater.*

There are other eaters of ants that are not considered members of the Order Edentata. These are the pangolins of the Order Pholidota, and the aardvarks of the Order Tubulidentata. The pangolins (fig. 13.17) live in Asia and Africa. Their most distinctive trait is a unique body cover-

FIG. 13.17 *Pangolin.*

ing of overlapping horny scales. The scales are flexible and sharp. When threatened the animal rolls itself into a ball, with the soft underparts completely protected by the armored back, legs, and tail. Aardvarks (fig. 13.18) live in South

FIG. 13.18 *Aardvark.*

Africa. They appear to have sprung from early plant-eating mammals (condylarths) rather than from insectivores. Aardvark is Afrikaans for "earth pig," a name which was suggested by the round blunt muzzle and piglike bristles of this strange animal.

Gnawers and Nibblers

We now turn to an exceptionally successful group of mammals, the rodents. Today rodents probably outnumber all other mammals combined and are found in every inhabitable land area of the world. They are the only placental mammal in Australia not introduced by man. Their adaptive radiation has produced arboreal forms like the squirrel, burrowers like the marmot, partially aquatic beavers and muskrat, and desert-dwelling jerboas and kangaroo rats. They are the dominant animals in the gnawing-nibbling category and have distinctive adaptations of teeth and jaws that reflect their specialization. Rodents lack canines and have only two prominent and opposing pairs of continuously growing incisors, one in the lower jaw and one in the upper. The outer surface of each incisor is made of enamel and is thus harder than the inner surface. In chewing, the incisors are sharpened as the harder enamel layer persists to form a chisel edge. A long gap, or diastema, lies between the chisel incisors and the cheek teeth. The latter are long, ridged, and efficient as grinders. The jaw muscles in rodents are large and the jaws have flat articulation so that they can grind fore and aft rather than up and down as in man.

Although fossils of rodents are known from strata as old as Paleocene, they are not common. Their small size makes them difficult to find, and searchers have a natural bias toward searching for larger and hence more spectacular fossils. Those that are found are often difficult to divide taxonomically. Divisions are based primarily on the disposition of the muscle scars (and in living rodents, the muscles themselves) which are associated with movement of the jaws. The protorogomorphs, which originated in the Paleocene, were ancestors of the later true squirrels and old world porcupines. They are represented in the fossil record by *Paramys* (fig. 13.19) which resembled a large squirrel, and by the unusual *Ceratogaulus,* the only rodent ever to have developed horns. Surviving protorogomorphs are represented by the small and thick-furred sewellels.

Squirrels are the characteristic sciuromorphs rodents, but this group also includes chipmunks and marmots. Beavers and their relatives are classified as castorimorphs. *Paleocastor* of the Pleistocene was a member of the beaver family that left casts of peculiar corkscrew burrows in Nebraskan fossil beds. The largest castorimorph was *Casteroides* of the Pleistocene which was about the size of a bear.

FIG. 13.19 Paramys.

The old world porcupines belong to the hystricomorph group, whereas new world porcupines, as well as capybaras, agoutis, chinchillas, and spiny rats are called caviomorphs. However, a group better known than these are the myomorphs which evolved from either sciuromorphs or hystricomorphs during the Oligocene. Those amiable pets, the hamsters (*Mesocricetus*), fall into this category along with rats, mice, dormice, pocket gophers, jerboas, and kangaroo rats. Among this inventory of rodents, rats and mice have formed a one-sided partnership with mankind. They invade his home, eat his food, convey some of his diseases, and voyage with him to the far corners of the world. Their persistence, tolerance, and tenacity are legendary.

Rabbits, hares, and pikas are called lagomorphs. For many years, they were classified as rodents, mainly because of their enlarged incisors. However, rodents have only a single pair of chisels above in opposition to a single pair below. Lagomorphs have two upper pairs and one lower pair. In addition, the cheek teeth are simpler and more numerous. They are elongate prisms with transverse ridges for cutting rather than the rodent manner of crushing. In the postcranial skeleton, the tail is reduced to a mere vestige, and hindlegs are strengthened and elongated for hopping. The lagomorphs are divided into two separate groups. On the one hand are the compact, short-legged pikas, or conys (fig. 13.20). These

FIG. 13.20 *Cony.*

animals lack the long ears of rabbits. Sometimes called mouse rabbits, the pikas are only about eight inches long. They are mountain dwellers and have been recorded on Mount Everest at elevations of seventeen thousand feet. In contrast to the pikas, hares and rabbits have developed into swift runners with long ears for acute reception of sound. The earliest rabbits are fossils of *Eurymylus* from the Paleocene of Mongolia.

Like the rodents, rabbits have been an annoyance to man because of the crops they consume and the diseases they carry. However they are used as pets and provide furs for clothing and meat for the table. As a result, they do not occasion as much ill-feeling as do rats and mice. Rabbits were introduced to Australia by man for his hunting pleasure, and, because of their fantastically rapid rate of reproduction, quickly populated the prairies, causing problems to ranchers and the native marsupials.

Flesh Eaters

Perhaps because man has evolved psychologically as a hunter, he has always been fascinated by the flesh-eating mammals. This is particularly true in regard to those carnivorous animals that, like man, form cohesive social groups. Indeed, a knowledge of the environmental and selective forces that shaped these animal societies has recently been shown to be of value in better understanding the origins of human society and behavior.

Life for the carnivore is in some ways more difficult than for the herbivore. Food for the plant eater is usually all about them, and thus they can be more relaxed about obtaining it. The flesh eater, however, has to track, attack, and overcome his dinner. Naturally, they must be faster and craftier than their prey. They require keen senses, especially the sense of smell, as well as body and limbs capable of agile, powerful movements. The weapons of carnivores are clawed feet, sharp teeth (especially effective canines), and powerfully muscled jaws for clamping onto their victims and tearing apart the flesh. The canines are used for stabbing, and cheek teeth called carnassials cut through meat and sinew like the blades of scissors.

Seventy million years ago in the Cretaceous Period, small carnivores of apparent insectivore ancestry began to appear. These archaic flesh eaters are called creodonts (fig. 13.21). Their overall appearance was not unlike that of weasels. The creodonts were relatively slow of wit and foot, but were able to bring down their similarly limited prey. These early predators had low, rather long skulls and smallish brains for mammals. The molar teeth were triangular and thus reminiscent of pantothere molars. There were no specialized carnassials. Although many of the creodonts remained small and unspecialized, some took on the general appearances of cats, dogs, bears, and hyaenas. *Hyaenodon* was a large carnivore with stout hyenalike teeth. This beast may very well have preyed upon the large titanotheres of the Oligocene.

FIG. 13.21 *Creodont.*

The creodonts are not considered ancestral to later groups of carnivorous mammals. In fact, the Order Creodonta probably arose independently from the stock that also produced the Order Carnivora. Having similar habits, members of the two orders often evolved similar appearances. It appears that with the coming of more advanced herbivores toward the end of Eocene time, the creodonts gradually lost ground to the carnivora. Early carnivora, particularly a group known as miacids, shared the predatory niche with Paleocene and Eocene Creodonta. In contrast to creodonts, the miacids had more anteriorly placed carnassials, as in many of their modern descendents. Larger brained than the creodonts, the miacids evolved rapidly in many adaptive directions, developing in the process a group of descendents called fissipeds. The fissipeds ultimately became the cats, dogs, weasels, raccoons, bears, and other land-dwelling flesh eaters of the modern world.

The dog family was doing well on earth long before man came along to make a companion of him. Very primitive canids like *Cynodictis* were hunting game over forty million years ago. The main line of dog evolution progressed from *Hesperocynon* of the Oligocene, to the Miocene *Cynodesmus* (fig. 13.22), to *Tomarctus* of the Pliocene, finally to Ice Age and modern dogs of the genus *Canis.* Dogs are fascinating animals. Wild dogs hunt by running their prey down. They require longer limbs, with ulna and radius locked to prevent rotation. The braincase is larger to accommodate elaboration of the cerebral cortex. Because dogs cannot truly outdistance most hoofed animals, they have developed the tactic of hunting in packs and employing ingenious strategems to exhaust and panic their prey. Wolves, for example, station members of the pack at various points along the route they plan to drive their prey and thus are able to take turns during the chase. Communication is essential in this breakneck pursuit of the fleeing dinner.

FIG. 13.22 Cynodesmas.

Not all canids live and hunt in packs. Coyotes, foxes, and jackals hunt singly or in pairs. Like all canids, they are alert animals that hunt either during the day or night, summer or winter, disdaining the habit of hibernation of many other mammals.

Whereas dogs are social animals, cats are known for their independence, stealth, and cunning. The most primitive members of the cat family are the civets (viverrids), which superficially resemble the older miacids. During the Pliocene, a side branch of the civets produced the hyenas, a group of scavengers not popularly associated with cats because of their somewhat doglike appearance. The true cats appeared in Late Eocene, and no noteworthy changes in their appearance has occurred since then. Padded feet with retractible claws are definitive features of true cats. They have only one molar in each tooth series, two of the premolars function as shearing teeth, and the anterior premolars are reduced or absent. Early in their evolution the cats developed along two separate lines; namely, the stabbing cats exemplified by the famous *Smilodon* (fig. 13.23) and the "biting cats," or true cats, typified by *Dinictis* (fig. 13.24). The former may have preyed upon some of the large herbivorous animals of the Pleistocene. Smaller brained, heavier, and slower than the true cats, their method probably was to leap upon their prey, grip the body with their powerful legs, and with mouth widely agape, repeatedly plunge the daggerlike canines into the body of the victim. The jaws

FIG. 13.23 *Skull of* Smilodon.

were articulated to permit their being opened widely, and nostrils were in a posterior position so that the animal could breathe while its fangs and face were pushed deep into the flesh of the victim. It has been suggested that wisdom was not one of the attributes of the stabbing cats. In the famous La Brea tar pit fossil site, there are about thirty stabbing cats to every true cat. The latter were apparently able to realize the dangers associated with the animals caught in the treacherous tar. Besides, true cats probably pursued the lighter and swifter herbivores of the plains. Like such living cats as the jaguars, pumas, and lynxes, brute force is no more important than craftiness, agility, and speed. Once within range, a mighty spring and they are upon their prey slashing away with rapid movements of the head and jaws.

Traits of aloofness, particularly among the uncommunicative house cat, have frequently been viewed as mysterious by humans. However, these personality traits are only the behavioral residue of the days in which cats were solitary hunters dependent on sagacious use of every faculty for survival.

In addition to the cats (felids) and dogs (canids), the land-dwelling carnivores also include raccoons, bears, weasels, and genets. Raccoons are omnivorous animals whose teeth have largely lost their shearing function. They are first recorded in the Miocene. The raccoon family (Procyonidae) includes the pandas of Asia, and the coatis and kinkajous of the Americas. Bears apparently evolved from bearlike dogs in the Miocene. Like the raccoons, they are omnivorous, have all of their toes retained and functional, and teeth that do not slice but grind. *Arctotherium* was an Ice Age bear, larger than any living member of the lineage.

Weasels, along with badgers, skunks, wolverines, and otters are mustelids. They are an interesting and varied group that first appeared about thirty-five million years ago in the guise of *Plesictes.* Living mustelids, like the martens and minks, are savage hunters, often killing

FIG. 13.24 Dinictis.

without reason. The genets, civets, and mongooses are viverrids, many of which also have bad-smelling glandular secretions which, paradoxically, are prized as a base for making expensive perfumes. The mongoose is well known as the champion snake killer of the animal kingdom, although some authorities question their reputation for killing vipers.

The carnivores of the Order Pinnipedia make their living in or near the sea. These are the seals, sea lions, and walruses. Pinnepeds adapted so quickly to an aquatic habitat that their ancestral pathway is obscure. Fossils older than Miocene are unknown. They have retained some doglike traits and may have descended from mammals resembling otters. Propulsion is accomplished by the limbs, which have been modified into paddles, and by sinuous movements of the body. The walrus and sea lion are capable of turning their rear limbs forward or backward and thus can use these limbs for moving about on land. Seals differ in that the hind limbs are fixed into a backward orientation. As a result, they must hobble along on their bellies when on land. Sea lions have small external ears, a trait which is absent both in walruses and seals. The teeth of pinnepeds readily reflect their dietary preferences. In addition to the obvious tusk canines of the walruses, they have broad crushing teeth to deal with the shells of oysters and clams. Seals and sea lions have pointed teeth for snaring fish.

Man it seems, takes pleasure in the company of the carnivores he uses as pets. He respects their intelligence, enjoys their loyalty, and admires their prowess in the hunt. Perhaps more than any other animal, carnivores help man to view himself in a more natural perspective.

Plant Eaters

Animals that walk upon hoofs and feed upon plants are loosely designated ungulates. It appears that the progression of events that produced ungulates began with primitive carnivores, proceeded to omnivorous or insectivorous animals, and then to "complete" herbivores. This is not an unexpected progression, for at least some descendents of former carnivores might be expected to take advantage of abundant and easily obtained plant food. However the shift to a different food does require certain modifications. Because plant food has a relatively low protein content, the plant eater must consume greater quantities of food per unit of body weight. Changes in the digestive tract are also required. Most important, dentition must be modified to deal with large amounts of very abrasive forage. The generalized low-crowned, sharp molars of the primitive placentals were obviously not suited for the task, and changes followed relatively rapidly. In general, incisors become more closely aligned into a curved arc for nipping. The canines were either lost or modified. Cheek teeth, which received plants cropped by the incisors, were variously modified to improve their efficiency in grinding. For example, the surface area was increased and altered to a squared-off outline. The once sharp tooth crests were rounded into low humps or complex ridges of folded enamel (fig. 13.25). In addition, the teeth developed high crowns, so that they stood considerably higher and could stand much more wear before being worn down to the roots. Premolars became enlarged and modified. In many ungulates they form a continuous series with the molars. The high-crowned cheek teeth are covered over with cement, filling the depressions between folds of enamel, so that sharp ridges of resistant enamel are maintained by the faster rate

FIG. 13.25 *Development of high-crowned grinding molars in horses. (From left to right,* Hyracotherium, Merychippus, *and* Equus. *Enamel shown in black.)*

of wear of the adjacent cement and dentine. For defense, some ungulates developed horns or antlers, while others grew enlarged canines or incisors. However, by far the more common defense was no defense at all, but speedy flight. To this end the limb and foot bones are lengthened, and all but the primitive or the heavy ungulates ran, like sprinters, upon their toes. Limb bones were further modified to prevent sprain producing rotation, and there is a tendency toward reduction in the number of toes. The hoofs are marvelous adaptations for protecting the toe bones from the repeated shocks of running over hard prairie sod.

The ungulates probably diverged from basic placental stock near the end of the Cretaceous. By Paleocene time the first ungulates appeared. They are named condylarths, and vaguely resembled some of the archaic carnivores. Near the end of the Paleocene, a condylarth appeared with squared-off cheek teeth and blunt, broad claws. The creature has been named *Tetraclaenodon,* and is the presumed ancestral type for the best known of the primitive ungulates, *Phenacodus* (fig. 13.26). With its large canine teeth and long catlike tail, *Phenacodus* did indeed look like a flesh eater. However, as

FIG. 13.26 Phenacodus.

in all true ungulates, a clavicle was lacking, and each of the five toes terminated in a small hoof rather than a claw. In general appearance, *Phenacodus* resembled a tapir, sans the tapir's proboscis. Like many of the Early Tertiary mammals, *Phenacodus* walked upon all five toes in a manner called plantigrade. Speed was not essential to this forest-dwelling browser.

FIG. 13.28 Uintatherium.

FIG. 13.27 Coryphodon, *a pantodont.*

At an early stage in their evolutionary history, the ungulates gave rise to rather bulky animals called amblypods. These creatures are represented by such extinct genera as *Coryphodon* (fig. 13.27) of the Lower Eocene and the elephantine, six-horned *Uintatherium* (fig. 13.28). These were ponderous, rather small-brained beasts. No doubt they provided an occasional good meal for the contemporary creodonts. The amblypods did not survive beyond Early Oligocene time. Unlike the condylarths, they were not ancestral to other more advanced lines of ungulates.

An interesting sidelight to the history of hoofed mammals is the development of a distinctive ungulate fauna in South America. In Paleocene time, some of the condylarths wandered southward across the Panamanian land bridge. Shortly thereafter the connection was severed, and the now isolated South American animals experienced their own distinctive and varied radiation. Included were creatures as small as mice and as large as elephants. Some had hoofs and some had claws. Most ate leaves or grass, but others developed rodentlike incisors and occupied the gnawing-nibbling niche. Litopterns evolved along pathways similar to horses, reducing the toes from five to three, and finally to one. Other litopterns like *Macrauchenia* (fig. 13.29) became somewhat camellike, with the added feature of a short, flexible trunk. Yet another group called notoungulates produced *Toxodon* (fig. 13.30), an animal as ponderous as a modern rhinoceros, and the smaller *Thomashuxleyi* (fig. 13.31). Mimicking the elephant, pyrotheres like *Pyrotherium* (fig. 13.32) developed massive bodies, short trunks, tusks, and large molars similar to those in more primitive mammals. All in all, these South American ungulates provide many striking examples of evolutionary convergence, as they followed adaptive trends resembling

FIG. 13.29 Macrauchenia, *a South American Litoptern.*

those of only distantly related herbivores elsewhere in the world. If the connection between the two continents had not been reestablished in the Plio-Pleistocene, these creatures (or their descendents) might still be prospering. However, the land bridge permitted migration of apparently somewhat superior North American animals. Perhaps because they were unable to stand the competition, the Litopterns and their contemporaries became extinct.

The modern and more familiar ungulates fall into two great orders; namely, the perissodactyls and artiodactyls. The perissodactyls include such modern grazers as the horse, tapir, rhinoceros; as

FIG. 13.30 Toxodon, *a South American Taxodont.*

FIG. 13.31 Thomashuxleya.

well as two extinct groups, the chalicotheres and titanotheres. They take their name from the Greek word *perissos*, which means extraordinary, and so in arithmetic refers to an odd number. Thus perissodactyls are odd-toed ungulates. The axis of the foot along which the weight of the body is supported lies through the third and middle toe. There is a tendency in perissodactyls for reduc-

FIG. 13.32 Pyrotherium.

tion of lateral toes from five to four, to a single toe as in the modern horse *Equus*. The sequence in the reduction can be demonstrated by placing the hand flat on the tabletop. Next, raise the wrist to simulate elevation of the foot for running on the toes. Note that the thumb leaves the table first, then the fifth finger, and finally, if the weight is restricted to the third finger, the second and fourth finger no longer touch the table.

The most primitive perissodactyl, *Hyracotherium* (fig. 13.33), appears very

FIG. 13.33 Hyracotherium.

near the close of the Paleocene Epoch. Familiarly known as "eohippus," this animal almost certainly evolved from a member of the condylarth group. *Hyracotherium* is important not only because it is so closely related to the ancestry of the perissodactyls, but also because it is the earliest member of the horse family. The animal was small, rather lightly built, with an arched back and slender legs. As an indication of trends that were to make greater speed possible, the ankles were elevated. Four tiny hoofed toes on the front feet, and three on the hind feet still touched the ground. The dentition was relatively primitive, with bluntly cusped, low-crowned molars, and premolars that were not yet shaped like the molars. These teeth were for browsing and quite suitable for plant eaters that lived in the forests rather than the plains.

From a base represented by eohippus, horses began to develop in an initially straightforward manner, with progressive increases in size, length of legs, width of incisors, height of crowns of molar teeth, size of braincase, and length of face. The middle toe received emphasis, as the lateral toes were reduced. There was a straightening and stiffening of the back. One can also discern a tendency for the premolars to become similar to the molars (a process termed molarization), and all cheek teeth develop increasingly complicated patterns of enamel ridges on the oral surfaces. *Miohippus* (fig. 13.34), an Oligocene horse, was about the

FIG. 13.34 Miohippus.

size of a sheep and had three toes on each foot. The central toe was larger than the two lateral toes.

After Oligocene time, the family tree of horses developed numerous side branches. By this time, large tracts of woodland were being replaced by broad open savannahs where grazing and running would be abetted. Although some horses remained primarily browsers, others quickly adapted themselves to the different food supply and the open environment. The Miocene horse, *Merychippus* (fig. 13.35) was in the vanguard of the new wave of plains-dwelling horses. This horse walked on a single

FIG. 13.35 Merychippus.

middle toe, the lateral toes being much reduced and of little use. The face was elongated and jaws already deepening to accommodate the lengthening molars. In the following epoch, two groups of horses arose from the general *Merychippus* stock. One group consisted of closely related genera typified by *Hipparion*. This was a fast-running type, with progressive skull and teeth, but conservative in the retention of three-toed feet. They migrated from North America into Europe and survived until earliest Pleistocene. The other group centered around *Pliohippus* and was somewhat more progressive. This horse was functionally single-toed, with vestiges of the side toes concealed beneath the skin of the upper part of the foot. The descendents of *Pliohippus* also branched into two groups. One of these is embodied in the genus *Hippidium* which migrated into South America when the land bridge was established. The other branch is represented by *Equus*, the genus to which all modern horses belong. *Equus* became widespread in America, Eurasia, and Africa, but suffered extinction in North America a few thousand years ago. The continent was restocked with the progeny of horses that had escaped from Spanish explorers in the sixteenth century. Thus North America, the birthplace of the horse lineage, became the first major continent to lose its horses. Some believe the cause of this extinction was disease. Others speculate it may have resulted from overkill by bands of prehistoric human hunters.

Horses, like dogs, have always had a close association with man. Prior to the invention of the internal combustion engine, they provided man with transportation, powered certain of his machines, pulled his heavy loads and plows, fought in wars, and even served as food. However, horses today are used primarily for sport, Hollywood westerns, and, alas, cigarette advertisements.

Tapirs (fig. 13.36) are a less prominent group of perissodactyls than horses. Mod-

FIG. 13.36 *Tapir.*

ern tapirs retain the primitive condition of four toes on the forefeet and three on the hind feet. Like the early horses, tapirs have low-crowned teeth which reflect their leaf-eating, forest-dwelling habits. The first true tapirs appeared in Oligocene time. Their evolution has been conservative, with the principal trends consisting of moderate increases in size and lengthening of the nose into a short flexible trunk. The latter trend is evidenced in fossil tapirs by the shortening of the nasal bones. Immature tapirs show light colored stripes and spots such as may have existed on early perissodactyls like *Hyracotherium*.

The remaining group of surviving perissodactyls are the rhinoceroses. They began their evolution during the Eocene as small, swift-running, hornless animals with four-toed front feet. During the following Oligocene Epoch, rhinoceroses appeared that were larger, heavier, and three-toed. These also were hornless. In the course of their history, rhinoceroses developed many variations. *Baluchitherium* (fig. 13.37) was a hornless giant that attained a height of over seventeen feet. Although still smaller than some of the dinosaurs, *Baluchitherium* appears to be the largest land mammal thus far discovered. Other rhinoceroses evolved into short-legged water dwellers, some remained horned inhabitants of tropical regions, and some became the hoary, hair-covered beasts of the ice ages. Modern rhinoceroses are entirely old world animals that appear to be the end members of a dying group. They include the black rhinoceros of Africa, the African white rhino, and the Asian two-horned rhino. A single-horned rhinoceros lives in Java, and another variant barely survives in India. The horns in living and fossil rhinoceroses consist neither of true "horn" nor of solid bone. It is composed of hair bundled and "glued" into a tough, compact mass. Rhinoceroses are favorite exhibits in zoos around the world. They seem to reflect antiquity and obsolescence.

FIG. 13.37 Baluchitherium.

There were a number of families of perissodactyls that did not survive into modern times. Among these were the titanotheres and chalicotheres. Like other perissodactyls, the titanotheres began their evolution with small, slender forms having four toes in front and three in the rear feet. The animal was only about eighteen inches long. From this petite beginning, the trend was toward an eventual dramatic increase in size. By Eocene time, *Eotitanops* (fig. 13.38) had reached the dimensions of sheep, and by Oligo-

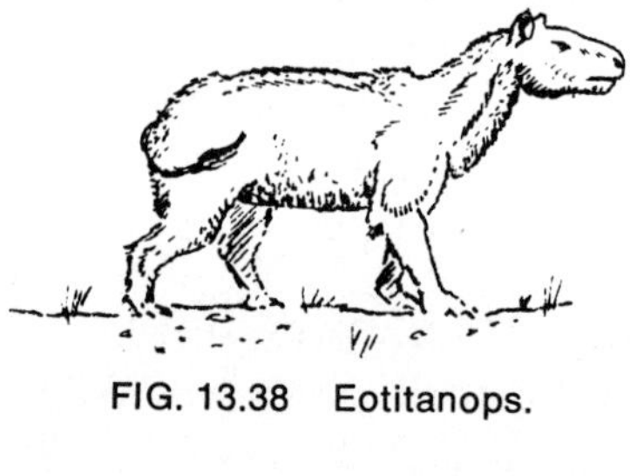

FIG. 13.38 Eotitanops.

cene the ponderous, horned *Brontotherium* had appeared (fig. 13.39). These massive creatures retained the toe elements of earlier titanotheres and developed thick elephantine limbs to support their great weight. The enormous head contained a brain no bigger than a baseball. Intelligence was probably not an area in which titanotheres excelled. Large numbers of titanotheres inhabited the western part of North America during the Oligocene, as is indicated by the abundance of their bones in beds of the South Dakota Badlands. *Brontotherium* is readily remembered because of the stout pair of hornlike processes that grew over the nose, and provided an effective bludgeon for disposal of enemies and sexual rivals. Other titanotheres did not develop horns at all, but enlarged their canines into effective weapons for defense.

FIG. 13.39 Brontotherium.

During most of their history, titanotheres lived in North America. By Early Oligocene some had migrated into Eurasia; however, they became extinct the world over by the middle of the Oligocene Epoch. Some believe that the cause of their relatively short tenure on earth was a developing scarcity in the kind of food they preferred. Basically browsers, the low-crowned teeth of titanotheres were probably poorly suited to the ubiquitous post-Oligocene diet of cereals and grasses.

The chalicotheres, the other extinct group of perissodactyls, have one truly bizarre characteristic. Unlike all other perissodactyls, they had three small claws, rather than hoofs, on their feet. Furthermore, the innermost, rather than the middle toe, was longest. These odd creatures were in many ways similar to horses in the appearance of the head and torso, yet their dentition resembled that of titanotheres. The North American chalicothere known as *Moropus* (fig. 13.40) was as large as a stallion, although generally more bulky. The front legs were longer than the rear, so that the back sloped rearward. Chalicotheres outlasted the titanotheres. They ranged from Eocene into the Pleistocene. Very likely they lived along the banks of streams using their clawed feet to loosen roots and tubers on which they fed.

A referee evaluating the evolutionary success of herbivores would have to rule that the odd-toed ungulates have not ultimately been successful. Chalicotheres and titanotheres are extinct, rhinoceroses and tapirs greatly reduced, and the horse family survives primarily because of the protection and control of man. More successful in terms of survival, variety, and numbers are the even-toed ungulates, or artiodactyls.

FIG. 13.40 *The chalicothere* Moropus. *(Adapted from a painting by Jay H. Matternes, from U.S. National Museum postcard.)*

Artiodactyls of today include pigs, hippos, deer, camels, goats, sheep, and cattle. Although it is true that they are the dominant ungulates of today, there has been a decrease in their numbers coincident with the expansion of man. Man has utilized artiodactyls throughout his history for meat, milk, hides, wool, and materials for weapons. Unlike perissodactyls which characteristically were centered in the Western Hemisphere, artiodactyls developed primarily in Europe, Asia, and Africa. They came on the scene during the Eocene, and by Late Tertiary had clearly achieved numerical and varietal superiority over the even-toed grazers.

Artiodactyls are characterized by an even number of toes—either two or four—with the axis of the foot trending between the third and fourth toes. The "big toe" and "thumb" are lost. In addition, one of the bones in the ankle, the astralagus, is distinctive. This bone forms the main articulation between leg and foot. In most nonartiodactyls it has a curved, pulley-shaped upper surface which provides free movement there, but is flat below. In artiodactyls, however, the lower as well as the upper end of the bone is pulley-shaped, and this provides greater freedom of motion in running. Another difference between perissodactyls and artiodactyls is in the dentition. Premolars and molars are dissimilar in artiodactyls, whereas in perissodactyls molarization of the premolars produces an entire series of similarly looking cheek teeth.

The earliest artiodactyls show little evolutionary progress beyond the condylarth stage, except for the changes in the astralagus mentioned above. The Late Eocene and Oligocene produced some interesting artiodactyls that became extinct before the Pliocene. Among these were large animals that bore a superficial resemblance to hogs (fig. 13.41). These bison-sized beasts, termed entelodonts, developed ugly flanges of bone along the sides of the skull and jaws. The result was, to our view, a decidedly repulsive face. Another extinct group were the oreodonts. Judging from the large numbers of oreodont bones found in strata of South Dakota, these animals roamed the western plains of North America in great herds. They are sometimes referred to as "ruminating swine" because their teeth indicate they were probably cud chewers, yet their short, stocky limbs give them a piglike aspect. Oreodonts are strictly North American fossils and are represented by peculiar clawed varieties like *Agriochoerus* (fig. 13.42) as well as the

FIG. 13.41 *Entelodont.*

perhaps more representative hoofed *Merycoidon* (fig. 13.43).

The differentiation of artiodactyls that began in the Eocene ultimately resulted in three major surviving groups represented by swine, camels, and ruminants such as deer and cattle. In many ways, the swine family is the most primitive of these groups. Even though the side toes are reduced, they are still four-toed ani-

FIG. 13.42 Agriochoerus.

mals that do not exhibit marked tendencies toward elongation of limbs. Swine are not specialized for a diet of cereals as are many more advanced artiodactyls. Indeed, they will consume nearly any food from snakes to squash with equal relish. The swine group includes both pigs and peccaries. Both began their evolution early in the Oligocene, but pigs are considered the more primitive of the two. They retain longer side toes, and

FIG. 13.43 Merycoidon.

these are on the ground and functional. Peccaries are functionally two-toed and have relatively longer legs and shorter, deeper skulls. Living pigs include the European wild boar from which our domestic pig is derived, as well as the unglamorous wart hog of Africa and the odd-looking babirusa with its curving tusks that grow upward over the sides of the skull. Today peccaries are found mainly in South America, although one species occurs as far north as Texas. Pigs are basically forest dwellers. They are considered very intelligent beasts, which may have contributed to their long survival.

Hippopotami are closely related to the Suina. They are the only ungulates to become amphibious. Fossil hippos are not known in rocks older than Pliocene. The hippopotami apparently arose from a group of Miocene piglike animals called anthracotheres. Today hippos live only in Africa. The larger species inhabit major streams where they eat water plants. They are well adapted for this life in the water. The small eyes are located high on top of the skull so that they can see while partially submerged. The nostrils, located on the top of the snout, can be closed to prevent shipping water when the animal is submerged.

Camels and llamas are closely related members of the Suborder Tylopoda. Although they are cud-chewing animals, they are not usually considered ruminants because of their long separate history, and the fact that they have a three-chambered, rather than a four-chambered, stomach. North America was the cradle of tylopod evolution. They first appeared in the Eocene as small animals not much bigger than a fox. As they evolved through the Cenozoic, they rapidly lost the side toes and most, but not all, increased the overall length of the legs and neck. Other changes included loss of two pairs of incisors in the upper jaws, leaving only the outer pair which bite down behind the lower incisors. In place of the missing incisors is a resistant pad against which the lower incisors grind. The molar teeth are high crowned and crested. In relatively recent time, foot pads for walking on sand developed, as well as the familiar hump for reserve storage of fat. Until only fifteen million years ago, all tylopods resembled llamas. The Pliocene form, *Alticamelus* (fig. 13.44), had a giraffelike appearance. Modern-looking llamas and camels were numerous in North America during the Pleistocene. The llamas moved southward and gained a homeland in the mountains and plains of South America. Camels migrated across the Bering Land Bridge to Eurasia and Africa where they soon established themselves as desert dwellers. Meanwhile, they mysteriously underwent extinction in their original North American homeland.

FIG. 13.44 Alticamelus.

Camels have a reputation for being ornery, perverse, and foul-smelling beasts. Nevertheless, they have been a boon to desert tribesmen. Well-adapted for a difficult environment, they can eat food other animals could not digest. They tolerate small amounts of salt water, if necessary, and are protected from blowing sand by nostrils that can be closed, hairy ears, and heavy eyelashes. You can tell when a camel is hungry, because as the fat is extracted from the hump, it shrinks and leans to the side. Living camels include the entirely domesticated, single-humped dromedary and the double-humped, wild bactrian camel of the Gobi Desert. Llamas have three living species; namely, the wild guanaco, the llama, and the alpaca.

We next come to the most varied and numerous of modern-day artiodactyls, the ruminants. Ruminants take their name from "rumen," the first of the four compartments of their compound, four-chambered stomach. Food is stored in the rumen where fermentation and bacterial action break down the difficult-to-digest cellulosic matter. It then passes into the second compartment, the reticulum. This is a honeycombed affair in which the food is formed into pellets that are then returned to the mouth for additional chewing. The rechewed food then passes to the third stomach, the psalterium, for further mashing and mixing, and finally to the last chamber, the abomasum, where it is combined with gastric juices, further reduced, and passed into the long digestive tract. The advantage of this method of processing is more than digestive efficiency. Ruminants in the wild state can crop large quantities of food while in unprotected open areas, and then retire to more secluded places for a relaxed period of cud-chewing and slow assimilation of food.

Ruminants can be divided into two groups, the tragulines and the pecorans. The tragulines are the least advanced. They have always lacked antlers, and some retained four toes. Tragulines are represented today by the timid and tiny chevrotains, or mouse deer, of the old world tropics. The ancestral traguline, *Archaeomeryx* (fig. 13.45) from the Upper Eocene of Mongolia, was a very similar animal.

FIG. 13.45 Archaeomeryx.

The pecorans take their name from the Latin word pecuniary, in reference to the Roman custom of measuring wealth by means of the number of cows they owned. In addition to cattle, percorans include sheep, deer, and giraffes.

The deer are a group that appear more clearly derived from the tragulids than the other ruminants. They have been primarily forest dwellers and browsers since they began their evolution in Oligocene time. *Blastomeryx*, which appeared about twenty million years ago in the Miocene, was more or less typical of early deer. Small in stature and lacking antlers, *Blastomeryx* resembled the modern Asiatic musk deer.

Antlers are the characteristic features of deer. They differ from horns in that horns are covered with a substance resembling that in claws and fingernails. Antlers are composed of bone and lack this horny covering. They also differ in that they are branched structures that are shed and renewed each year. Antlers are really periodic outgrowths of the bones located above the eye orbits. While growing, the antler is covered with a velvet skin. After the antlers have reached full size, the skin dries, and the deer rubs it off in preparation for the sexual combat and display that occurs during the mating season. Antlers, both in ancient and modern deer, come in scores of sizes and shapes. The record in largeness goes to *Megaloceros* (fig. 13.46) whose antlers measured eleven feet from tip to tip. It is possible these great structures were used, not only in actual sexual combat, but as a psychological deterent to combat, as other males measured their chances against the opposing buck.

The deer expanded across North America, Europe, and even migrated into South America during the Pleistocene. Although a few can be found today in northwestern Africa, they never achieved success in that great continent, possibly because of its prior occupation by the antelopes.

Giraffes are rather progressive ruminants that probably branched off from the deer family during the Miocene. They specialized in eating the leaves of trees (especially acacia) that grew in areas where most other herbivores were grazers. Long legs and a long neck give giraffes an obvious advantage in obtaining food that grows high off the ground, and also helps in early detection of approaching carnivores. *Palaeotragus,* the ancestral giraffe, was an animal closely similar to the living okapi of the Congo. *Palaeotragus* had a sloping back and two smallish skin-covered "horns" suggestive of similar structures on modern giraffes. The neck was not yet unduly long. From such a beginning, the giraffes developed along two lines. The first was toward massive oxlike beasts characterized best by *Sivatherium.* The second trend emphasized leg and neck elongation and gave rise to the modern giraffe. Remarkably, the increase in neck length has not involved the addition of a single vertebra. Like mammals generally, the giraffe has seven cervical vertebrae, although these are greatly elongated. Giraffes are gentle animals that protect themselves by running and by use of keen senses. When cornered, however, they can use their hoofs to lethal advantage.

FIG. 13.46 Megaloceros.

The most numerous of the higher ruminants are bovoids. This group includes the cattle, sheep, and goats on which man has come to depend. Less familiar bovoids are muskoxen, yaks, pronghorn antelope, kudu, and gnus, as well as water buffaloes, bison, gazelles, wildebeests, and impalas. The first bovoids appeared during the Miocene and, although relative latecomers, soon became the dominant hoofed mammals of the world. Some became heavily built two-toed animals with high-crowned teeth for grinding grasses. True horns are the distinctive trademark of the bovoids. They are generally permanent and occur in both sexes. One exception is the North American prongbuck which rather resembles an antelope more than a bovoid. The prongbuck does not shed the bone core of his forked horn, but does drop the horneous covering. Some taxonomists prefer to place them in a separate family from the Bovidae.

Domestication of bovoids has been persistent for many centuries. Domestic cattle are believed to have been derived from the extinct Late Pleistocene aurochs. However, even before the domestication of cattle, men kept and used sheep and goats as they traveled across Asia more than nine thousand years ago.

Mixed Company

One can hardly imagine more outwardly dissimilar but related animals than sea cows, elephants, and conies. Yet the early representatives of each of these groups show similarities which suggest a common African origin. The conies (Hyracoidea) are small rodentlike animals that date back to about thirty-five million years ago. They are now found in the lands which border the Mediterranean. Conies have chisel-shaped incisors, and the molars show a pattern not unlike that seen in rhinoceroses. There are four toes on the front feet and three on the rear. Special foot pads assist these quick-moving little plant eaters in scampering about on rocky ledges, and even tree limbs.

Another animal that may have shared a common ancestor with the elephants lived in Egypt more than thirty million years ago. This Oligocene animal has been named *Arsinoitherium* (fig. 13.47) and was as large as a rhinoceros. Of the four horns on its head, two were massive bony affairs that arose from the top of the huge skull. Little is known about the predecessors of *Arsinoitherium*, nor have any descendents as yet been discovered. They stand alone at a point in time, providing a challenge to bone hunters.

It is hard to imagine an animal with such incongruous traits as to resemble a sea cow and an elephant at the same time. Such animals lived on both sides of the Pacific Ocean during the Middle and Late Tertiary. They are called desmostylids. *Desmostylus* has the elongate skull of the mastodon, tusks, and a short trunk. Yet the animal was aquatic and fed on molluscs.

The dugong and manatees are relatives of the elephants that have adapted so completely to an aquatic mode of life that the hindlegs are lost. There is a well-developed tail flap that provides propulsion, and the front limbs are modified into flippers. Steller's sea cow was a dugong that measured twenty feet or more in length and lived in the icy waters

FIG. 13.47 Arsinoitherium.

of the Bering Strait until exterminated by man. Sea cows and manatees (fig. 13.48) are termed Sirenia, in reference to the beautiful sirens of Greek mythology. However, it is unlikely that a close view of a real sirenian would evoke ecstasy in the heart of any Greek oarsman. They

FIG. 13.48 *Manatee.*

are interesting but graceless creatures that browse for aquatic plants along coastal areas. The dugongs inhabit the shores of the Indian Ocean and Red Sea, whereas manatees prefer the shores of the tropical Atlantic. Manatees can be distinguished from dugongs by their rounded, rather than notched, tail and the deeper split in their upper lip.

Sirenians began their evolution early in the Tertiary and quickly spread about the shorelines of the world. The earliest fossils come from the Eocene of Egypt, and these bones show clearly the features that link the sea cows to the elephants and conies.

Among the very large mammals still living today, elephants seem unique in nobility of behavior and physical majesty. Man has been consistently fond of these colossal animals. Because they have a trunk, elephants and their older relatives, the mastodons and mammoths, are called proboscideans. Elephants are hoofed animals which carry their great weight on pillarlike legs and cushioned, five-toed feet. Other animals as tall as elephants compensated for the distance to a food source near the ground by developing longer necks. However, proboscideans have a thick, short neck to support their massive heads. They use their trunks for getting food and water, and for other purposes as well. Modern members of the proboscidean lineage have developed short, high skulls which provide more efficient leverage for lifting with trunk and tusks. In elephants, the walls of the skull are greatly thickened with frothy cellular bone called diploe, thereby increasing leverage further and providing more area for the attachment of neck muscles and ligaments without greatly increasing the weight of the skull. The great trunk is formed of the elongated upper lip and nose. The nasal passages extend through the trunk and are encased by over five hundred muscles that provide both great power and dexterity. The trunk may terminate in either a single fingerlike process as in the Indian elephant, or in two processes as in the African elephant. These terminal processes are useful in grasping small objects, including the peanuts we feed the circus elephants. In fossil proboscideans, the gradual development of the trunk can be followed by noting the position of the external nares on the skull, for these openings recede far up on the skull in sequential stages of trunk development.

Although less famous than the trunk, the elephant's teeth are also distinctive. The tusks, for example, evolved from the second pair of incisors, and they continue to grow throughout the life of the elephant. Furthermore, there are only four or six fully formed teeth at any one time in the jaws of the adult elephant. These grinders have the familiar complex ridges of enamel seen in other grazers. The cheek teeth are formed successively in the rear of the jaws and move into position to replace worn predecessors.

Moeritherium (fig. 13.49) is the generic name for the ancestral proboscidean.

FIG. 13.49 Moeritherium.

Fossils of moeritheres have been recovered from Lower Eocene and Early Oligocene beds near ancient Lake Moeris some sixty miles southwest of Cairo, Egypt. These early members of the proboscidean lineage were probably tapirlike in general size and form. The nares on the front of the skull give no evidence of a trunk, but there may have been a pendant lip. The molars are quadrate, and have four blunt cusps which are connected to form two transverse ridges. The second incisors of both upper and lower jaws are enlarged as the beginning of the trend toward tusk development.

The ancestry of the moeritheres is not definitely known. It is likely that they were derived from Early Tertiary condylarths. However, the post-moerithere history is better known, as proboscideans diverge into one branch that leads toward mastodons and elephants, and another that terminates with the extinct dinotheres.

The link between the dinotheres and moeritheres is inferred, for the intermediate forms have not been discovered. Dinotheres were large African and Eurasian animals that survived from Late Miocene into Pliocene time. They were characterized by tusks that curved downward and backward from the lower jaws. No doubt this unusual tusk orientation was useful in uprooting plants and digging for roots and tubers. There were no tusks in the upper jaws. Each jaw was equipped with five, low-crowned cheek teeth, each having two or three cross ridges. Dinotheres had long heavy legs and, like modern elephants, well-developed trunks.

The lineage of proboscideans that was to culminate in today's elephants led from moeritheres to primitive Oligocene mastodons like *Palaeomastodon* and *Phomia* (fig. 13.50). The skulls in these

FIG. 13.50 Phomia.

early forms were larger and higher in the vertical plane than in moeritheres, and the nostrils had receded backward indicating the presence of a proboscis. The body and legs were also larger than in *Moeritherium*. Tusks were present in both jaws in these animals.

There is a gap in the fossil record between *Phomia* of the Early Oligocene, and the long-jawed Late Miocene mastodon known as *Gomphotherium* (fig. 13.51). *Gomphotherium* exceeded his predecessors both in overall body size and in the length of the jaws. The trunk was also well developed. Like all morphological features, the elephant's trunk appeared gradually through generations of evolving pachyderms. For those who prefer a more fanciful explanation, there is Kipling's tale of how the small wriggly nose of the elephant child was seized by a crocodile on the banks of the Limpopo River and stretched to the length at which now we find it.

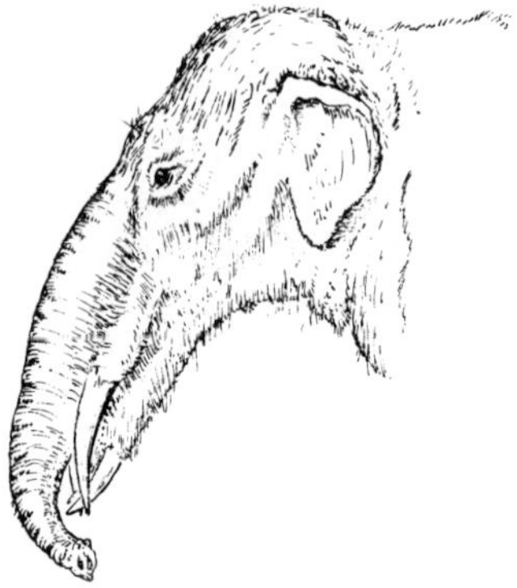

FIG. 13.51 Gomphotherium.

There were many interesting side branches of the elephant family tree. In one group typified by *Stegomastodon*, the lower tusks were suppressed and the lower jaws shortened. The upper tusks became large upturned structures resembling tusks in modern elephants, except that these animals were really " long-jawed mastadons" and not elephants at all. Other long-jawed mastodons had the forward part of the lower jaws and tusks turned downward, as in *Rhynchotherium*. The shovel-tuskers, such as *Amebelodon* of the Pliocene, had broad, scoop-shaped tusks which were used for digging up vegetation.

While on the one hand, the long-jawed mastodons were evolving, there was another separate line of mastodons with more familiar short jaws. The lengthy evolution of this group culminated in the well-known Ice Age mastodon *Mammut*. Remains of this great beast are found throughout North America. A large femur, embellished with a bright red ribbon, was displayed at a Missouri county fair several years ago. It was facetiously labeled, "Boone County Ham Bone." *Mammut* was heavier, but not as tall as modern African elephants. It had the sharply crested molars of short-jawed mastodons, but a more elongated skull, and large, strongly curved tusks. Some of the long, reddish hair of *Mammut* has been found unaltered. Carbon-14 dating techniques indicate *Mammut* roamed North America as recently as 8,000 years ago and may have been hunted by paleoindians.

Elephants evolved from mastodons. Their line of progression is traced backward to a Miocene animal named *Stegolophodon*. This creature was rather a cross between a long-jawed mastodon on the one hand, and an archaic elephant on the other. Next on the evolutionary ladder was *Stegodon* of the Pliocene. *Stegodon* was decidedly elephantine in its short, tuskless lower jaw. The molar teeth were elongated and developed numerous cross ridges as a prelude to the even more complicated system of ridges characteristic of mammoths and elephants. In *Stegodon* we first see the development of the peculiar elephant solution to the problem of giant molars in shortened jaws. There is only one tooth in each jaw at any given time. Other teeth are formed behind and above the tooth sockets in the upper jaw, and behind and beneath the sockets in the lower jaw. As the newly formed teeth come in and push forward, the worn teeth in front are shoved outward until they finally break off and are lost. The deep crowns and multiplication of enamel ridges develop quickly in Pleistocene elephants. These teeth were suitable for a diet of tough grasses. They are usually found more deeply worn than the teeth with cone-like cusps that identify the mastodons. Less wear resulted from the crushing of the less abrasive foliage on which mastodons fed.

Ice Age elephants of North America, Eurasia, and Africa are commonly referred to as mammoths. There have been a variety of species of mammoths. A woodland form in Europe and Asia evolved straight tusks and reached heights of over fourteen feet at the shoulder. The great "imperial elephant" had tusks measuring thirteen to sixteen feet on the curve. The columbian elephant had immense spiral tusks that overlapped at their tips. However, not all species were large. Pigmy elephants less than six feet tall once lived on the isolated islands of Malta and Cyprus. Certainly, the most famous of all the extinct elephants was the wooly mammoth. The beast is known to us from many bone sites, and even from frozen carcasses found in Siberia and Alaska. This was the majestic animal seen alive by our own ancestors and depicted with admirable fidelity on the walls of cave dwellings. Ancient Indians probably hunted these animals along with mastodons. Then about eight thousand years ago, all but two genera of proboscideans became extinct. The survivors are *Elephas,* the Asiastic elephant, and *Loxodonta* of Africa. The larger size and great fanlike ears of the African elephant distinguish it readily from the Asian form. The domed forehead of *Elephas* is yet another distinguishing trait.

Judging from the former great variety and number of proboscideans, it is clear they are now on the road to extinction. Perhaps they may survive if people do better protecting them. In an age in which man fears for his own survival, the decimated remnants of these grand animals should act as reminders of the vulnerability of all creatures.

SUMMARY

Mammals evolved from reptilian ancestors during the Triassic Period. The earliest of these mammals were small, mouselike creatures. Throughout the remaining periods of the Mesozoic, mammals remained relatively small and inconspicuous. However, the widespread extinctions of the great reptile groups in the Late Cretaceous set the stage for rapid expansion of the mammals during the Cenozoic.

Among the attributes of mammals are various natural devices for keeping the body at a constant temperature, thus making activity possible even in cold climates or colder parts of the day. This "warm-bloodedness" is accomplished principally by an insulating cover of fur or hair and an efficient circulatory system. Blood in mammals is pumped by a four-chambered heart which is really a dual structure, with the right side pumping blood only from the body to the lungs, and the left side from the lungs to the body.

Additional mammalian traits are glands for suckling the offspring, a single bone (the dentary) in the lower jaw, the presence of malleus and incus bones to accompany the stapes in the auditory apparatus, relatively larger brain, double occipital condyle at the base of the skull, greater differentiation of teeth, ribless lumbar region, and positioning of the legs more directly under the body.

Living mammals are grouped into three categories: placentals, marsupials, and monotremes. These groups differ in details of reproduction. All have internal fertilization and nourish the young with milk after birth. However, the monotremes lay reptilelike, rather large eggs which are incubated. Marsupials have minute eggs which then develop for a short period of time in the uterus. The immature "embryo" then crawls into the pouch on the abdomen of the mother and attaches itself to one of the mammary nipples. In placental mammals, the tiny eggs are retained for a longer period in the uterus, which becomes modified for their reception and nourishment.

Features such as the inflected angle of the lower jaw, syndactylous digits, and the triangular patterns of molar teeth are used in identifying the remains of fossil marsupials. The marsupials and placentals probably evolved from common ancestry during the middle part of the Cretaceous. During the Cenozoic, placental mammals expanded more rapidly than marsupials. The two island continents where marsupials fared best were South America and Australia.

The early mammals of the Mesozoic Era are distinguished by their teeth. They include the docodonts with irregularly quadrate molars, triconodonts which have three cusps in a row on molars, multituberculates having longitudinal rows of low cusps on elongate molars, and the symmetrodonts and pantotheres with more or less triangular cusp arrangements on the cheek teeth.

Shrewlike insectivores approximate the appearance of the first placental mammals. From such creatures, either directly or indirectly, the higher orders of placental mammals expanded. In the vanguard were small-brained, archaic ungulates such as amblypods and condylarths, primitive flesh-eating creodonts, and the prosimian beginnings of the primate line. A second wave of placentals began their evolution early in the Middle Cenozoic, and most of the modern orders became established. Their adaptive radiation was extensive. Bats evolved as the only mammals to have achieved true flight. Some mammals returned to the sea (whales, seals, porpoises). Others on land fed upon the flesh of plant-eating hoofed animals that were simultaneously undergoing diversification. Members of the Order Carnivora replaced the earlier creodonts and produced canids, cats, raccoons, bears, seals, and many others. The beginnings of modern hoofed animals occurred in Late Paleocene and Eocene, and they soon divided into the perissodactyls and artiodactyls. Among the perissodactyls, the evolution of the horse family is particularly well known, leading by various branches from diminutive *Hyracotherium* of the Late Paleocene to the modern horse *Equus.* Other perissodactyls include rhinoceroses and tapirs, as well as the extinct chalicotheres and titanotheres. No less important as herbivores are the artiodactyls. Late Eocene and Oligocene artiodactyls include the hoglike entelodonts and the numerous oreodonts. Surviving groups include swine, camels, hippopotami, and such ruminants as cattle, deer, giraffes, and tragulines. As the large herbivores evolved during the second half of the Cenozoic, they experienced specialization of teeth and limbs for grazing. Grass tends to wear down grinding teeth. To compensate for the wear, grass eaters have high-crowned teeth that grow through most of their lives.

The earliest proboscideans (elephants and their extinct relatives the mammoths and mastodons) have been found as fossils in Eocene rocks of North Africa. These early forms were smaller and had shorter trunks than modern elephants. Thus proboscidean evolution is marked by an increase in size, lengthening of the trunk, and changes in tooth and skull structure.

SUPPLEMENTARY READINGS AND REFERENCES

Carrington, Richard (and editors of *Life*). 1963. *The mammals,* Life Nature Library. New York: Time Inc.

Moody, Paul A. 1962. *Introduction to evolution.* New York: Harper.

Osborn, Henry F. 1910. *The age of mammals in Europe, Asia, and North America.* New York: Macmillan.

Romer, A. S. 1971. *Vertebrate paleontology.* 3d ed. Chicago: Univ. of Chicago Press.

Scott, Wm. B. 1937. *A history of land mammals in the Western Hemisphere.* New York: Macmillan.

Simpson, G. G. 1945. The principles of classification and a classfication of mammals. *Bull. Amer. Mus. Nat. Hist.* 85: 350 pp.

———. 1951. *Horses.* London: Oxford University Press.

Watson, D. M. S. 1946. The evoluton of proboscidea. *Biol. Revue* 21: 15–29.

Young, J. Z. 1957. *The life of mammals.* London: Oxford University Press.

Zappler, Georg, and Zappler, Lisbeth. 1970. *The world after the dinosaurs.* Garden City, New York: Natural History Press.

14 The Manipulators

Man is a fraction of the animal world. Our history is an afterthought, no more, tacked to an infinite calendar. We are not so unique as we should like to believe. And if man in time of need seeks deeper knowledge of himself, then he must explore those animal horizons from which we have made our quick little march.

—Robert Ardrey, *African Genesis*

One might make a convincing argument against considering primates separately from other mammals as has been done here. To do so might imply that primates are special, unusually successful, or the ultimate goal of vertebrate evolution. This is certainly not the case. Evolution has no preselected goals. The sparrow and the opossum have been just as successful in their own mode of life as the monkey and the ape. It is not because of a sense of snobbish superiority that we deal with the primates last, but because they are our closest relatives, and therefore interest us keenly.

AMONG THE TREES

The Order Primates has considerable variety. It can be divided into the Suborder Prosimii which includes the tree shrews, lemurs, lorises, and tarsiers; and the Suborder Anthropoidea which contains monkeys, apes, and man.

Throughout their history primates have retained generalized and rather primitive traits. It has been stated that the very key to their survival has been adaptive flexibility and lack of advanced specializations. For example, they retain the primitive number of five fingers and toes. The teeth are less specialized than in many carnivores and herbivores, and they grow neither horns, hoofs, trunks, or antlers. This does not mean that they did not respond adaptively to their environment, for life in the trees led to many changes important in the eventual evolution of man.

One of the most significant of these changes was the development of the grasping hand with its opposable thumb. This feature not only permitted the primate a firm grip on his perch, but allowed him to grasp, release, manipulate, and control a variety of objects in his leafy home. The forearms retained their primitive mobility in that the radius and ulna were able to rotate on one another permitting the hands to be reversed in position. Claws, no longer as necessary for climbing, were eventually replaced by fingernails. The development of the grasping hand was accompanied by another important trend. For an animal that leaps from branch to branch and seeks its food from lofty vantage points, certain visual attributes would be advantageous. In the course of evolution the eyes of primates became positioned toward the front of the face, so that there was considerable overlap of both fields of vision. Because of this evolved trait, primates are able to judge with survival accuracy depths, distances, and heights. They are able to assess possibilities for lifesaving alternatives. Improved binocular vision was only a component of the eye-brain-hand system that would be utilized by later primates to change the very face of the earth.

Still other changes in primates were secondarily related to changes in the eyes and limbs. To protect the eye from bulging jaw muscles, a postorbital bar developed behind the eye orbits. As the sense of smell diminished in importance, the snout was reduced so that the face became flatter. In response to the habit of swinging from branches, forelimbs and hindlimbs diverged in form and function and a general predisposition toward upright posture developed that proceeded naturally toward bipedalism in man.

PREMONKEYS

The living tree shrew of Borneo represents the approximate stage of organization from which the higher primates evolved. They are so primitive that some zoologists question their membership in the primate order. Nevertheless, the animal has many skeletal similarities to early fossil prosimians, including very mobile and versatile fingers.

The fossil record for the primates begins with two species of a creature named *Purgatorius* recently discovered in the Hell Creek Formation of Purgatory Hill in Montana. These finds offer evidence that primates emerged from insectivore ancestry late in the age of dinosaurs. Unfortunately, the remains of *Purgatorius* consist only of a few teeth. *Plesiadapis* (fig. 14.1) of the Paleocene of Europe and North America has a much better fossil record. The pattern of the crowns on the cheek teeth and the structure of the limb bones indicate that *Plesiadapis* was an early lemur. However, the incisors were decidedly rodentlike and, as in rodents, were separated by a diastema from the

FIG. 14.1 Plesiadapis, *a Paleocene prosimian. (Modified from E. L. Simons, 1964, "Early Relatives of Man,"* Scientific American *211:56.)*

premolars. Thus plesiadapoids were small gnawing animals which may ultimately have been pressed into extinction by the ascendency of the rodents. If this is true, man may owe the rat and his kin a debt of gratitude for forcing the directions of primate evolution away from the gnawer and nibbler niche.

Succeeding steps in the early evolution of primates involved trends toward increased binocularity, brain size, and snout reduction. *Notharctus* (fig. 14.2) was a true lemur that spread through the tropics of Eocene North America into Europe and clearly displays these trends.

FIG. 14.2 Notharctus.

With the coming of cooler climates in the Oligocene, prosimians appear to have been forced southward into the southern parts of Asia and into Africa as well. The island of Madagascar has been their chief refuge. It was apparently separated from Africa after the arrival of lemuroids and lacked carnivores and competitive primates that might decimate prosimian populations.

Living lemurs are an interesting and unusual group. They include the aye-aye which has rodentlike incisors and an exceptionally long third finger used for digging insects from the bark of trees, and the larger indri with its hairless face, hands, and feet. An offshoot of the lemurs are the lorises of Africa and Asia. Lorises have more closely spaced eyes than lemurs, are nocturnal, and have the bizarre habit of sometimes hanging upside down while feeding.

In the Late Paleocene and Eocene, the prosimians known as tarsiers made their appearance. *Tetonius* (fig. 14.3) was a typical Eocene tarsier in which some of the characteristics of modern forms were already evident. The muzzle that characterized the lemuroids had disappeared, and the large eye orbits were oriented directly forward. The braincase was also expanded. Like the lemurs, tarsiers were widespread on the northern continents in the tropical Early Tertiary, but later moved southward in response to the cooling trend experienced at higher latitudes. Today the surviving genus is *Tarsius* (fig. 14.4) of the East Indies and Philippines. They are specialized animals with nocturnal habits, enormous closely spaced eyes, naked ears and tail, and heads which can swivel through an angle of nearly 360 degrees.

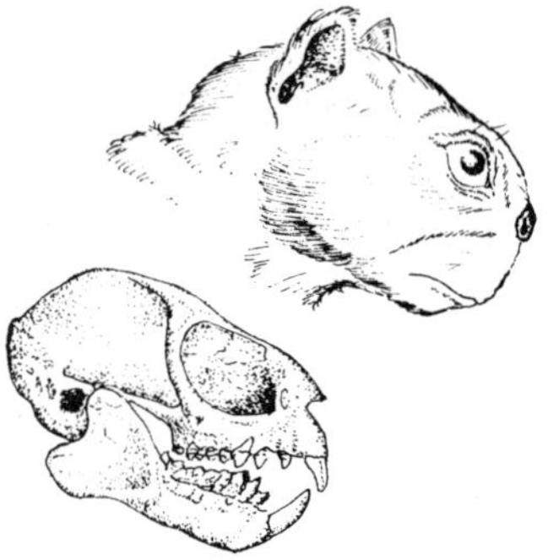

FIG. 14.3 Tetonius.

FIG. 14.4 *Tarsier.*

MONKEYS AND APES

It seems likely that the higher primates, the Anthropoidea, were derived from animals at the approximate structural level of the tarsiers. The anthropoids include three large superfamilies; the Ceboidea, or "new world monkeys," the Cercopithecoidea, or "old world monkeys," and the Hominoidea. This last superfamily is further divided into the Family Hylobatidae (which includes gibbons), Pongidae (orangutans, chimpanzees, and gorillas) and Hominidae (the family of man). Monkeys represent the next step in the evolution above prosimian levels. Their progress is evident in their forward facing eyes, more complex molars, larger braincase, the boney bar protecting the eye orbit, and improved hands. The new world group, although once more widespread, are now found only in South America. They differ from their old world cousins in having an extra premolar and the possession of a prehensile tail. Fossils of new world monkeys are rare, consisting of a few fragments of bone and teeth from Columbia and Argentina. However, they are represented by twelve living genera, and these include spider monkeys, howlers, marmosets, and capuchins. The latter are well known to children as the companion of the organ grinder.

The record for monkeys begins with a group of Oligocene fossils from the arid Fayum district of Egypt. Three groups appear to be represented by these bones. The first is a ceboidlike family represented by *Parapithecus* (fig. 14.5); the remaining two groups have definite old world characteristics. Included here is a gibbonlike group represented by *Aeolopithecus, Aegyptopithecus,* and *Propliopithecus.* E. L. Simons (1964) has proposed that *Propliopithecus* may have been along the evolutionary pathway that led toward pongids and hominids.

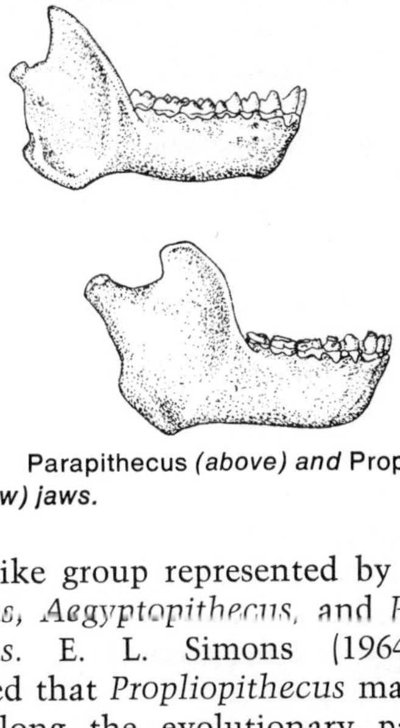

FIG. 14.5 Parapithecus *(above) and* Propliopithecus *(below) jaws.*

Today the more advanced old world monkeys are widely distributed in Asia and Africa. They include such familiar animals as the baboons, langurs, guenons, and the numerous macaques, or rhesus monkeys. There are about a dozen species of rhesus monkeys. Notable among these are the so-called Barbary "apes," the only monkeys now living in Europe. Legend has it that when they are no longer residents of the Rock of Gibralter, the British Empire will have expired. Rhesus monkeys are sacred animals in India and important experimental animals in biomedical laboratories. Indeed, it was the versatile little rhesus that made the first trip aboard a spacecraft.

The Miocene Epoch saw the continuation of the trend toward the hominid stage. *Pliopithecus* (fig. 14.6) lived about twenty million years ago and shows many subtle advances beyond the gibbon level. The forelimbs and hindlimbs in *Pliopithecus* are of equal length, and thus differed from the gibbon which has longer arms than legs. Still further advances are seen in a group of lightly built apes named dryopithecines that roamed Southern Europe, Asia, and Africa in Miocene and Pliocene time. Dryopithecines apparently lived predominantly on the ground in open savannah country. Their dental arcade was beginning to show an arch similar in outline to the human condition, although the projection of the canines was more apelike. Many paleontologists feel that, although the dryopithecines were close to the common stock of man and apes, they had already embarked on the branch leading to apes rather than man.

Modern apes belong to the family Pongidae, and include the gibbon, orangutan, chimpanzees, and gorillas. They live in the tropics of Africa and Asia, are intelligent, lack tails, are able to distinguish colors, and can make a variety of sounds and expressions to indicate their feelings and reactions. Chimpanzees are, next to man, the wisest animals alive. They fashion and use simple tools, live and hunt in social groups, and care for their young for as long as eight years. These intangible traits probably also existed in the dryopithecines, and thus chimpanzee behavior has always been intriguing to physical anthropologists and phychologists.

FIG. 14.6 Pliopithecus.

THE FAMILY OF MAN

Recently it has been demonstrated that, among the Mio-Pliocene dryopithecines, there emerged a hominid which can be considered a direct ancestor of modern man. This first hominid is named *Ramapithecus*. The genus is known primarily from jaw fragments discovered in India and also in Kenya. The adult animal was probably no larger than a chimpanzee. Although the fossil fragments are scant, it is evident that the face was foreshortened, and the teeth had many similarities with those of man. In contrast to the condition in apes, the incisors were smaller, the canines not much larger than premolars, and the dental arcade quite humanlike. Lacking even large canines for defense, these lithe, vulnerable protohumans may very well have defended themselves with sticks, stones, or even cudgels made of the limb bones of larger animals.

The story of the origin of modern man began to be more fully understood about a half century ago with the discovery of a juvenile "ape-man" that was given the name *Australopithecus africans* (fig. 14.7). Since that time, many additional skulls and skeletal parts have been found in Late Pliocene to Mid-Pleistocene beds of southern and eastern Africa, as well as Israel, Java, and China. They are not at all a homogeneous group and bear several distinct generic names. In the interest of simplicity, the group is colloquially termed "the australopithecines."

Australopithecines include two rather separate lines of evolving hominids. The first is characterized by larger-sized animals than the other. The smaller types, like *Australopithecus africans,* were lightly built bipedal primates with man-like teeth, but brains much smaller than modern man's brain. It is generally held that these animals fed primarily upon small animals that they either killed themselves or scavenged. The larger and somewhat heavier australopithecines are typified by *Paranthropus.* They were also small brained, but their dentition indicates they fed upon plants. Both large and small australopithecines possessed the sloping brow and receding lower jaw characteristic of early hominids. Brain capacity averaged about 500 cubic centimeters and was thus comparable to that in modern apes.

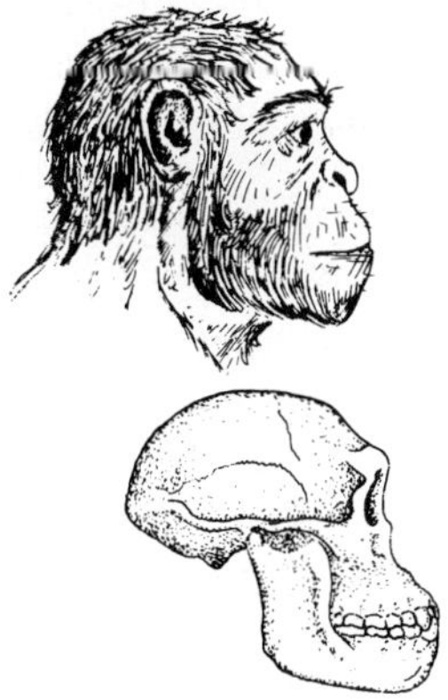

FIG. 14.7 Australopithecus.

As more and more australopithecine fossils are found, it becomes increasingly evident that the group was quite varied, and individual discoveries will continue to cause often acrimonious taxonomic debate. Some branches evidently made crude tools of stone ,whereas others only used objects already formed by nature. As we have seen, some preferred a diet of meat, others plants, and still others were probably omnivorous. Some like *Australopithecus bosei* (once called *Zinjanthropus*), *Australopithecus robustus,* and *Australopithecus africans* may represent side branches of the hominid tree that terminated in the Middle Pleistocene. The most advanced australopithecines may represent the true human stock. For example, *Homo habilis,* discovered at Oldovai Gorge, is considered by some anthropologists to be a direct ancestor of modern man. In defense of his view that these bones came from a primate above the evolutionary level of other australopithecines, L. S. B. Leakey noted the somewhat greater cranial capacity (about 680 cc), slightly less robust jaw, and modern aspect to the bones of the foot. Terminal bones of the thumb and index finger suggest an increased level of facility in handling and possibly working with objects. These attributes make *Homo habilis* an interesting stage between the general australopithecines on the one hand, and the succeeding pithecanthropines (*Homo erectus*) on the other.

PROMISE IN GRAY

With the advent of creatures like *Homo habilis,* the human lineage began to experience successive structural changes that set it more visibly apart from the other primates. The first and most important of these changes involved increases in the size and complexity of the brain. Man is "the thinking animal," and brain power, with its enormous adaptive advantages, has been the most decisive force in human evolution. In order to house this expanding organ, the capacity of the braincase was increased. The largest of living apes have a cranial capacity of about 500 or 600 cubic centimeters. The more primitive variants of *Homo erectus* (fig. 14.8) had a skull that would contain approximately 900 cubic centimeters of brain matter.

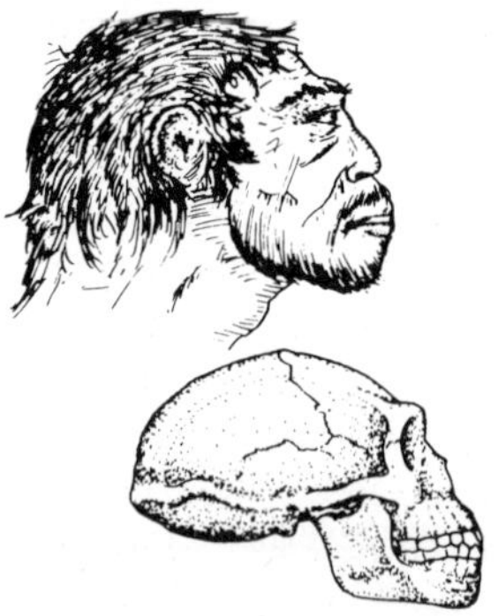

FIG. 14.8 Homo erectus.

In species of *Homo* from the Late Pleistocene and Recent, brain size has ranged from 1200 to over 2000 cubic centimeters. It seems likely that the use put to this great bulge of gray matter may have been increased as a consequence of the trend toward more fully erect posture. The forward arch of the back as seen in apes is gradually transformed into the S-shaped curve that characterizes modern man. The pelvis is swung into alignment, and the skull with its precious cargo becomes better balanced against gravity atop the spinal column. If only that cranium could tell us more of the intellectual capabilities of early men, or of the dim beginnings of speech and symbolic thought. On these subjects the bones are mute. We have only secondary indications of the state of intelligence from the tools and artifacts left along the way by a developing mankind.

So fine a brain requires slow and careful nurturing. This is accomplished in part by an impressive increase in the length of time it takes for human young to reach maturity. A long period of slow development preceding birth minimizes environmentally induced hazards and variation. It provides the time needed for the growth of complex neurological systems. After birth, care of young leads to the transmission of learning, which, although also found in other animals, reaches an unprecedented level in man. The development of family groups was probably a natural outgrowth of the lengthy episode of postnatal care in which survival of offspring required a cooperative division of tasks. From these family groups of prehistoric men, it was a logical step toward clan, tribe, and ultimately nation.

THE EMERGENCE OF MODERN MAN

The australopithecines were succeeded during the middle years of the great Ice Age by the larger-brained pithecanthropines. As was the case with their predecessors, pithecanthropines have been given many different names, but recent works suggest they can be included within the single species *Homo erectus*. Specimens of *Homo erectus* have been found in Africa, Asia, China, Indonesia, and Europe. In their physical characteristics, they nicely bridge the gap between modern man and the australopithecines. The similarity of pithecanthropine leg bones to those of present-day man is apparent. Cranial capacity ranged from 775 cubic centimeters to 1100 cubic centimeters, and thus overlaps at least the lowermost range of modern man. To accommodate the larger brain, the vault of the cranium is enlarged. In pithecanthropines, the skull was rather flattened, the jaws still jutted forward at the tooth line, heavy ridges persisted above the eyes, and a definite jutting chin was absent. These were primitive traits, but the dental arcade and teeth were essentially modern. Ample evidence indicates that *Homo erectus* made good use of his mental powers. He had discovered the use of fire, learned to make crude axes and scrapers, and may have indulged in religious rites that involved cannibalism.

The pithecanthropines were men of the Middle Pleistocene. They were followed in the Late Pleistocene by men of our own species, *Homo sapiens*. The more ancient of these peoples were first discovered in the Neander Valley near Dusseldorf, Germany. As a result, they have come to be known as Neanderthal men (fig. 14.9). With his heavy brow ridge, and prognathous and chinless jaw, Neanderthal man has become the very personification of the "caveman." His massive face seems a brutish carry-over from the Middle Pleistocene. Yet in other ways, Neanderthaloids were quite modern. Below the neck their skeleton matched our own, and their brain capacity equalled or exceeded that of present humans. Perhaps they had a slightly greater ruggedness of joints and muscles. Neanderthaloids knew the use of fire, cooperatively hunted large Pleistocene mammals, and buried their dead with an apparent belief in either afterlife or spiritual immortality. Their intellectual attributes were probably in no way inferior to the earliest of the men called Cro-Magnon that replaced them approximately thirty-two thousand years ago.

FIG. 14.9 *Neanderthal man.*

Cro-Magnon man was perhaps slightly less robust, had a more vertical brow, and a decidedly modern point to the chin (fig. 14.10). Five successive cultural stages of Cro-Magnon have been discovered. In the earlier of these stages, their implements were not unlike those of the Neanderthals (from whom they may have learned). However, they soon became more skillful artisans, producing points,

FIG. 14.10 *Cro-Magnon man.*

knives, scrapers, and gouging tools from flint; and harpoons, awls, and needles from bone. Burial of the dead became a more elaborate affair. Cro-Magnon art, displayed in carvings and cave drawings of animals and fertility venuses, is remarkable for its ethereal quality and elegant symmetry. Truly, the age of the philosopher had arrived.

Through most of man's early history, he was a wandering hunter or gatherer of wild edible plants. However, about seven thousand years ago, some human societies had begun to domesticate animals and plants so that permanent settlements could be developed. The results were pervasive changes in the way men lived. For the first time there were large and reliable sources of food. This not only made rapid increases in population possible, but brought food surplus, so that some individuals were spared the demands of providing food and could develop themselves into scholars, tradesmen, and priests. Indeed, the eventual rise of civilizations and cities was dependent upon these surpluses. Soon men were clearing woodlands to make room for crops and livestock, damming streams for irrigation, and changing the natural environment in ever increasing numbers of ways. For the first time in his two million year history, man became an enormously significant force in modifying the face of the planet. Simultaneously he was creating a kind of evolution the earth had not previously seen. It is called cultural evolution.

Cultural evolution is, of course, nongenetic. It expands and changes as a result of interthinking rather than interbreeding. It is born of ideas that were initially transmitted by gesture, grunt, and facial expression, and much later by speech and crude symbols scratched into clay and stone. Down the long corridors of human development, each generation of men has built upon the conceptions of their ancestors. It was done at an ever-increasing rate. Today this has resulted in what can be termed the technology explosion. Unfortunately the phenomenon has not brought with it a corresponding eruption of mass wisdom. The problems of too many humans, too much waste and pollution, and limited resources have become a persisting nightmare for all thinking persons. In the brief chapters of this book we have noted the lengthy tenure of other vertebrate orders. We humans would aspire to similar longevity. Should not the days of man be at least as numerous as those of saurians? After all, man is profoundly more adaptable, immensely more intelligent, and capable—if he wishes—of controlling even his own evolution. In the long history of animal life on earth, he is the only species with that particular capability. Besides, in the four to five hundred million years of the vertebrate annals, he has only just arrived. How can mankind provide for a secure and extended future? An easy answer to that question does not exist. Certainly survival will require a level of responsible, unselfish, and harmonious cooperation among all of *Homo sapiens* the world round. It will demand a vast overhauling of many of our most cherished beliefs. It will require that we change our very nature.

SUMMARY

The primate order can be divided into two suborders: the Prosimii which include lemurs, tarsiers, and tree shrews, and the Anthropoidea in which are classified monkeys, apes, and man. Among the general traits of primates are a lack of overspecialization, hands and feet with the primitive number of five flexible digits suited for grasping, fingernails rather than claws, eyes relatively close together for binocular vision, a tendency toward shortening of the snout, and an increase in the size of the brain. Precise vision and hands suitable for manipulating objects have been important factors in the evolution of the higher primates, including man.

The fossil record for primates begins among the prosimians with fragmentary remains of *Purgatorius* from Late Cretaceous strata. Other fossil prosimians include the rodentlike *Plesiadapis,* the lemur *Notharctus,* and a tarsier named *Tetonius.* During the middle part of the Cenozoic, the Ceboidea (new world monkeys), Cercopithecoidea (old world monkeys), and Hominoidea (gibbons, orangutans, chimpanzees, apes, and man) produced early representatives.

The group of rather variable Middle Cenozoic hominoids known as dryopithecines appear to have been close to the stock from which both man and ape arose. It is surmised that these small animals (less than a meter tall) inhabited open savannah country, for their posture was more erect than that of their monkey relatives. They possessed a tendency for the kind of rounding of the dental arcade that is seen in humans, but also had apelike projecting canines. *Ramapithecus,* a fossil hominid discovered in India (and later Africa), displays human tendencies in the shape of the jaw, the presence of five-cusped molars, reduction of canines, and foreshortening of the face. Hominids like *Ramapithecus* probably gave rise to the australopithecines.

Dating studies of australopithecines reveal a chronological range for these creatures of from one-half to five million years old. They were a varied group, species of which combined both ape and human traits. Cranial capacity was about 600 cubic centimeters (about the same as in chimpanzees). Although some apelike traits persisted, the dentition resembled that in man, they were more upright in posture than their predecessors, and they appear to have made crude tools.

The next stage of human evolution is recognized by the appearance of *Homo erectus* during the Middle Pleistocene. These erect walking primitive men had a brain capacity of about 900 to 1100 cubic centimeters. They made tools such as stone axes equipped with handles and knew how to use fire.

Homo sapiens neanderthalensis made his debut about 90,000 years ago. His remains are widely distributed in Europe, Asia, and Africa. Modern man—*Homo sapiens sapiens*—appeared in the second interglacial stage of the Pleistocene. They left behind such evidence of cultural development as carefully fashioned tools, carvings, and cave paintings.

Along these pathways, modern man came into existence. After only a few tens of thousands of years he has reached a crisis. His further existence is threatened by overpopulation, dwindling resources, lack of food, and nuclear war.

SUPPLEMENTARY READINGS AND REFERENCES

Clark, W. E. Le Gros. 1967. *Man-apes or apemen?* New York: Holt, Rinehart and Winston.

Howell, W. W. 1966. Homo erectus. *Scientific American* 215 (5): 46–53.

Leakey, L. S. B. 1961. *The progress and evolution of man in Africa.* London.

Napier, John. 1962. The evolution of the hand. *Scientific American* 207 (6): 56–62.

———. 1967. The antiquity of human walking. *Scientific American* 216 (4): 56–66.

Pilbeam, D. 1972. *The ascent of man.* New York: Macmillan.

Simons, E. L. 1964. The early relatives of man. *Scientific American* 211 (1): 51–62.

Simpson, G. G. 1966. The biological nature of man. *Science* 152: 472–478.

Swanson, Carl P. 1973. *The natural history of man.* Englewood Cliffs, N.J.: Prentice-Hall.

Tobias, P. V. 1965. Early man in East Africa. *Science* 149: 22–33.

Van Valen, Lee, and Sloan, R. E. 1965. The earliest primates. *Science* 150: 743–744.

Weckler, J. E. 1957. Neanderthal man. *Scientific American* 197 (6): 89–96.

Appendix
A Catalogue of Organisms

This simplified classification includes all of the major phyla of organisms whether or not they are discussed in the text. Those categories which are extinct are indicated in italics.

PLANTS

Phylum Schizomycophyta: bacteria, actinomycetids
Phylum Myxomycophyta: slime molds
Phylum Euglenophyta: plantlike flagellates
Phylum Pyrrophyta: dinoflagellates
Phylum Cyanophyta: blue-green algae
Phylum Chlorophyta: green algae
Phylum Charophyta: stoneworts
Phylum Phaeophyta: brown algae
Phylum Rhodophyta: red algae
Phylum Chrysophyta: golden-brown algae, diatoms, coccolithophoridids
Phylum Mycophyta: fungi
Phylum Bryophyta: mosses and liverworts
Phylum Tracheophyta: vascular plants
- *Subphylum Psilopsida: psilopsids*
- Subphylum Lycopsida: lycopsids
- Subphylum Sphenopsida: sphenopsids
- Subphylum Pteropsida: ferns
- Subphylum Spermopsida: seed plants
  - *Class Pteridospermophyta:* seed ferns
  - Class Cycadophyta: cycads
  - Class Ginkgophyta: ginkgoes
  - Class Coniferophyta: conifers
  - Class Angiospermophyta: flowering plants

ANIMAL-LIKE PROTISTA OR "PROTOZOANS"

Phylum Sarcodina: amoebas, foraminiferans, radiolarians
Phylum Mastigophora: animal-like flagellates
Phylum Sporozoa: sporozoans
Phylum Ciliophora: ciliates like *Paramecium*

ANIMALS

Phylum Archaeocyatha: primitive spongelike organisms
Phylum Porifera: sponges
- Class Desmospongea: sponges with spongin and siliceous spicules
- Class Hyalospongea: sponges with siliceous spicules
- Class Calcispongea: sponges with calcareous spicules
- Class Sclerospongiae: sponges with aragonite and organic fibers

Phylum Coelenterata: jellyfishes, sea anemones, corals
- Class Scyphozoa: jellyfish
- Class Hydrozoa: hydrozoans
- Class Anthozoa: corals

Phylum Platyhelminthes: flatworms

Phylum Nemertinea: ribbonworms

Phylum Nemathelminthes: roundworms

Phylum Gordiacea: horsehair worms

Phylum Acanthocephala: spiny-headed worms

Phylum Trochelminthes: rotifers

Phylum Chaetognatha: arrowworms

Phylum Bryozoa: moss animals or bryozoans

Phylum Brachiopoda: brachiopods (lamp shells)

Phylum Mollusca: molluscs
- Class Gastropoda: snails
- Class Bivalvia (Pelecypoda): clams
- Class Cephalopoda: cephalopods
- Class Monoplacophora: primitive molluscs
- Class Amphineura: chitons
- Class Scaphopoda: tusk-shells

Phylum Annelida: segmented worms

Phylum Arthropoda: arthropods
- *Class Trilobita:* trilobites
- Class Crustacea: crustaceans
- Class Arachnida: spiders
- Class Insecta: insects

Phylum Echinodermata: echinoderms (starfish, etc.)
- *Class Edrioasteroidea: edrioasteroids*
- *Class Cystoidea: cystoids*
- *Class Blastoidea: blastoids*
- Class Crinoidea: crinoids
- Class Asteroidea: starfishes
- Class Ophiuroidea: serpent stars
- Class Echinoidea: echinoids (sea urchins)

Phylum Protochordata: acorn worms, pterobranchs
- Class Enteropneusta: acorn worms
- Class Pterobranchia: pterobranchs
- *Class Graptolithina: graptolites*

Phylum Chordata: chordate animals
- Subphylum Tunicata: tunicates
- Subphylum Cephalochordata: lancelets
- Subphylum Vertebrata: vertebrates
  - Class Agnatha: jawless fishes
  - *Class Acanthodii:* primitive jawed spiny fishes
  - *Class Placodermi:* primitive jawed fishes
  - Class Chondrichthyes: sharks, rays
  - Class Osteichthyes: bony fish
  - Class Amphibia: amphibians
  - Class Reptilia: reptiles
  - Class Aves: birds
  - Class Mammalia: mammals

Glossary of Selected Terms

Adaptation Any modification of an organism that better fits it for existence in its present environment or enables it to live in a different environment.

Adaptive Radiation The diversity that arises among species as each adapts to a unique set of environmental conditions.

Algae Any member of a group of simple plants usually classified in the thallophyta. Algae contain chlorophyll and are capable of photosynthesis.

Allele One of two or more contrasting genes that determine alternative characters in inheritance because they are situated at the same location on homologous chromosomes.

Amblypods Primitive Early Tertiary ungulates.

Amino Acids Any of a large group of nitrogenous organic compounds that serve as structural units in the proteins.

Ammonoid Cephalopods An extinct group of marine molluscs typically with coiled, chambered shell and septa with crenulated edges.

Amniotic Egg The type of egg found in reptiles, birds, and monotremes. The developing embryo is maintained and protected by means of an elaborate arrangement of shell membranes, yolk, sac, amnion, and allantois.

Amphibians "Cold-blooded" vertebrates that respire by gills in their early life stages and lungs as adults

Amphistylic A type of jaw suspension in which the palatoquadrate articulates with the bones of the braincase.

Anaerobic Organism An organism that does not require free oxygen for respiration, but makes use of such processes as fermentation to obtain its energy.

Angiosperms An advanced group of plants having floral reproductive structures and seeds in a closed ovary; the "flowering plants."

Anthozoa The corals.

Anthracosaurs Late Paleozoic labyrinthodont amphibians that evolved from ichthyostegids.

Anthropoidea The taxa which includes monkeys, apes, and men.

Antiarchs Small, often armored, placoderm "fishes" with jointed, moveable pectoral spines.

Anus The opening through which the solid refuse of digestion is excreted.

Aragonite A form of $CaCO_3$ which crystallizes in the orthorhombic system. The iridescent, inner "mother of pearl" layer in some shells and pearls is composed of aragonite.

Archaeocyatha A phylum of extinct marine organisms having double, perforated, calcareous, conical walls. Geologic range: Lower to Upper Cambrian.

Archosaurs Advanced diapsid reptiles, including the thecodonts, "dinosaurs," pterosaurs, and crocodilians.

Arthrodires A group of placoderm fish characterized by armor and jointed necks.

Articular Bone in the posterior portion of the lower jaw which forms the surface for the joint (L. *articulus,* a little joint).

Artiodactyls Hoofed mammals that typically have two or four toes on each foot.

Aspidin A hard, bony tissue comprising the armor of ostracoderms, and characterized by the absence of bone-cell spaces. Aspidin is considered to be the precursor of bone by some paleobiologists.

Autotroph An organism that uses an external source of energy to produce organic nutrients from simple inorganic chemicals.

Basalt A fine-grained igneous rock dominated by dark-colored minerals and consisting of more than 50 percent plagioclase feldspars and the balance ferromagnesium silicates.

Blastula A stage in embryological development where a single layer of cells surrounds a central cavity.

Bone A type of supportive tissue which functions in supporting the body and is characterized by its hard matrix impregnated with calcium and phosphorus salts.

Brachiating Swinging from branch to branch and tree to tree by use of the limbs, as exemplified by monkeys.

Brachiopod A type of bivalved marine invertebrate most abundant during the Paleozoic Era.

Bulla The tympanic bone surrounding the middle ear, resembling a swollen flask without a base.

Canids General term for members of the Family Canidae which includes dogs, wolves, foxes, and their relatives.

Canine Teeth The teeth located between the premolars and incisors in mammals. They are often somewhat longer than adjacent teeth.

Calcareous Composed entirely or in part of limy material such as calcium carbonate ($CaCO_3$).

Carnassial Teeth Cheek teeth in flesh-eating mammals that are modified for shearing and slicing flesh (L. *carnem,* flesh).

Carnivore An animal that feeds upon other animals.

Cartilage A type of supportive tissue consisting of large oval cells, each with a nucleus, and surrounded by a material called chondrin (a mixture of proteins with high percentage of collagen).

Cast The impression extracted from a mold.

Caudal Pertaining to the region of the tail (L. *cauda,* tail).

Cephalon (as in trilobites) The anterior body segment forming the "head," and consisting of several fused segments.

Ceratopsians The ornithischian horned "dinosaurs."

Cervical Pertaining to the neck region of an organism (L. *cervix,* neck).

Chalk A soft, white, fine-grained variety of limestone composed largely of the calcium carbonate skeletal remains of marine microfossils.

Chelonians Turtles and tortoises.

Chert A dense, hard rock, composed of submicrocrystalline quartz (SiO_2).

Chitin A resilient, tough organic compound secreted as exoskeleton material in many invertebrates, especially arthropods.

Choanocytes Special cells typically lining the internal cavity of a sponge, characterized by a delicate collar of protoplasm through which passes the lower part of a long flagellum.

Chlorophyll Any of several green, light-absorbing pigments essential to the process of photosynthesis. Found in green plants and certain algae and bacteria.

Chromosomes Microscopic, threadlike bodies composed of chromatin that appear in the nucleus of the cell at the time of cell division. Chromosomes contain the genes, and are normally constant in number for any particular species.

Cilia Short hairlike processes in some types of animal cells. Cilia are capable of beating in unison to produce locomotion or to move particles.

Civets Relatively primitive, long-bodied, long-tailed members of the Superfamily Feloidea which includes cats and hyenas as well as civets.

Cloaca A posterior region of the alimentary canal of certain animals into which the products from the intestines, kidneys, and sex organs are discharged.

Cnidoblast Special cells in coelenterates occurring most abundantly on the tentacles, which contain nematocysts, or stinging capsules. The extruded parts of nematocysts can cause paralysis in any minute animal they penetrate.

Coacervate Clusters of colloidal particles created when colloids of opposite electrical charge bump into each other and combine their "shells" into a single tightly bound molecular shell of water.

Coccolithophorids Planktonic, marine, biflagellate golden-brown algae which typically secrete a covering of discoidal calcareous elements called coccoliths.

Coelenterata The phylum of invertebrate animals that are characterized by a hollow body cavity, radial symmetry, and stinging cells; includes jellyfish, corals, and sea anemones.

Colloid Any finely divided substance in permanent suspension in a medium.

Conch Term typically applied to the shell of cephalopods and gastropods.

Condylarth The most primitive of the extinct archaic ungulates .

Conodonts Small toothlike fossils found in rocks ranging in age from Ordovician to Triassic and possibly derived from early vertebrates.

Continental Accretion A hypothesis stipulating that continents are enlarged by the formation of successive mobile belts at their margins.

Cosmic Dust Accumulations in outer space composed of tiny particles of frozen water, methane, ammonia, carbon dioxide; elementary permanent gases including hydrogen, and minute grains or crystals of mineral matter and metals.

Cosmoid Scales Scales consisting of two layers; a dentinelike cosmine layer with branching tubules underlain by a bony layer with a pulplike region between these layers.

Cotylosaur The earliest and most primitive reptiles.

Creodonts Early, primitive flesh-eating mammals.

Crossing Over An exchange of material between adjacent chromatids of an homologous pair of chromosomes; crossing over is caused by a chromosome break and produces genetic combinations different from those of the parents.

Cryptozoic The geologic time which preceded Paleozoic Era.

Cycloid Scale Nearly circular fish scales with concentric rings about a central point such as those found in the more primitive teleosts.

Ctenoid Scale Fish scales with concentric growth rings as in cycloid scales, and with small spines which extend outward on that portion of the scale not covered by the neighboring scales.

Density Mass per unit volume.

Dental Arcade The arc formed by the teeth when viewing the oral surface.

Dental Formula Notation scheme for designatng sequence and number of teeth in each side of upper and lower jaws. For example, 2–1–2–3 means each side of a jaw contains 2 incisors, 1 canine, 2 premolars, and 3 molars. (May also be expressed $\frac{2\text{–}1\text{–}2\text{–}3}{2\text{–}1\text{–}2\text{–}3}$ for upper + lower jaws on one side.)

Deoxyribonucleic Acid (DNA) A nucleic acid that is found chiefly in the nucleus of cells and functions in the transfer of genetic characteristics and in protein synthesis.

Dinoflagellates Unicellular marine algae usually having two flagella and a cellulose wall.

Dipleurula A hypothetical generalized larval form that follows the gastrula in the development of echinoderms. The larva can be pictured as an elongated, pear-shaped organism with mouth on the slightly flattened ventral side extending to an anus at the posterior end.

Diploid Having two sets of chromosomes forming pairs; as in somatic cells.

Dolostone A rock composed of the mineral dolomite, which is chemically a calcium magnesium carbonate.

Dromedary One of the two modern species of camel. The dromedary camel has one hump, the Bactrian camel has two.

Echinodermata A phylum of marine echinoderms, typically with spiny calcareous exoskeletons and five-fold radial symmetry.

Edentates An order of mammals that includes living armadillos, tree sloths, and South American anteaters.

Enzyme Any of various complex organic compounds originating from living cells and capable of producing certain chemical changes in organic substances by catalytic action.

Epiblast The outer "layer" of cells of the gastrula.

Epipubic Bones Bones in the pelvic structure of marsupials which function in the support of the pouch.

Epoch A division of geologic time; a subdivision of a geologic period.

Era A major division of geologic time which can be divided into geologic periods.

Eucaryote A type of living cell characterized by a true nucleus enclosed within a nuclear membrane, and by well-defined chromosomes, and organelles.

Euryapsid Pertaining to a group of extinct marine reptiles with temporal opening of the skull in a dorsal position above the postorbital and squamosal bones.

Evolution (organic) The process of descent with modification as a result of which present-day species of animals have arisen from those of the past.

Evolutionary Convergence The process by which similarity of form or structure arises among different organisms as a result of adaptation to similar habitats.

Femur The first or proximal long bone of the pelvic limb. The thigh bone.

Fermentation The partial enzymatic breakdown of organic fuels by an organism in the absence of oxygen. Fermentation can be termed anaerobic respiration.

Fibula One of the two long bones located below the femur in the pelvic limbs of most land vertebrates.

Filter Feeder An animal that obtains its food, usually in small particles, by filtering it from water.

Flagellum A whip-like protoplasmic process of a cell.

Foramen Magnum The opening at the base of the cranium through which the spinal cord joins to the brain.

Foraminiferida An order of mostly marine, unicellular animals (sarcodinids) that secrete tests usually composed of calcium carbonate.

Fossil The remains or evidence of the existence or nature of animals or plants that lived in the geologic past.

Gamete Either of two cells (male or female) which must unite in sexual reproduction to initiate the development of a new individual. Gametes are haploid.

Ganoid Scale A type of scale found in early actinopterygian fish consisting of outer layers of dense enamellike ganoin separated from an inner bony layer by vascular tissue.

Gametophyte The sexual, haploid generation of a plant that produces gametes. The gametophyte generation alternates with a diploid, asexual, sporophyte generation.

Ganglion	A mass of nerve cells.
Gastrula	An embryo in the two layer stage, consisting of ectoderm and endoderm enclosing the archenteron.
Gene	The unit of heredity transmitted in the chromosome.
Gene Pool	The total aggregate of allelic forms of all the genes in a population of organisms.
Genotype	The assortment of genes that make up the basic hereditary constitution of an individual.
Genus	The major subdivision of a family or subfamily of plants or animals, usually consisting of more than one species.
Geosyncline	A great structural trough in the earth's crust that receives a thick sequence of sediments over long spans of geologic time and may ultimately be compressed into a major system of mountains.
Gill Slits	A series of paired openings from the sides of the pharynx through the body wall to the exterior, found at some stage in all chordates.
Girdle	The skeletal structure of a vertebrate, lying approximately transverse to the backbone, upon which the paired fins or limbs are articulated.
Gonad	The essential reproductive organs of multicellular animals in which the gametes are formed.
Gondwanaland	The name of a hypothetical southern continental mass which fragmented into smaller segments comprising Africa, India, South America, and Australia.
Granite	A coarse to medium crystalline rock composed of potassium feldspars, quartz, and small amounts of ferromagnesian silicates.
Graptolites	Extinct, colonial, marine invertebrates ranging from the Late Cambrian to the Mississippian, and believed to be protochordates.
Gymnosperms	An informal term referring generally to seed plants in which the seeds are not enclosed ("naked seeds"), as in cycads and conifers.
Hadrosaurs	Ornithiscian "duckbill dinosaurs" of the Cretaceous.
Haemal Arches	Bony or cartilaginous arches formed on the ventral side of vertebrae below the centrum where transverse processes unite to enclose the caudal artery and vein.
Haploid	Having a single set of chromosomes to each cell or individual, as in gametes.
Herbivore	An animal that feeds upon plants.

Heterocercal Tail In fishes, a tail in which both lobes are different in size. More commonly the upper lobe is larger, the reverse being designated "reverse heterocercal."

Heterotroph An organism that depends on an external source of organic substances for its food and energy.

Heterozygous Referring to a pair of genes responsible for a particular characteristic, which are not alike.

Hexacorals Those corals which have six primary pairs of mesenteries. Those hexacorals lacking skeletons comprise the sea anemones, whereas those with skeletons are stony corals or scleractinids.

Hominoidea The Superfamily of the Suborder Anthropoidea which includes the families Hylobatidae (gibbons, siamangs), Pongidae (orangutans, chimpanzees, gorillas), and Hominidae (man).

Homology Correspondence in fundamental structure. Homologous organs are those similar in origin and in basic structure.

Homozygous Referring to a pair of genes responsible for a particular structure which are alike.

Humerus The uppermost, long bone of the forelimb of tetrapods. Normally the humerus articulates with the pectoral girdle above, and the ulna and radius below.

Hyomandibular The modified upper bone of the hyoid arch which functions as a connecting element between the jaws and the braincase in certain fishes.

Hyoid Arch In early jawed fishes, the first gill arch behind the jaws.

Hypoblast The inner layer of cells comprising the wall of the cavity or archenteron in the gastrula.

Hystricomorphs The Suborder of the Order Rodentia which includes the Old World porcupines.

Ictidosaur Therapsid or mammallike reptiles considered very close to mammals in skeletal characteristics.

Ilium The uppermost of the three bones located on each side of the pelvic girdle.

Incisors Teeth located at the front of the jaw, and usually modified for biting.

Insectivore An insectivorous animal is one which feeds upon insects. The Order Insectivora comprises a diverse group of small primitive placental mammals with five clawed toes, and elongated snouts. The order includes shrews and hedgehogs.

Intercentrum (of vertebra) An element of the interlocking vertebral structure of many Paleozoic amphibians, located beneath the neural arch.

Ions Atoms that have an electrical charge, resulting from the loss or gain of one or more electrons.

Ischium The most posterior of the three bones located on each side of the pelvic girdle.

Isotopes Atoms of an element that have the same number of protons in the nucleus, the same atomic number, and the same chemical properties, but which have different atomic masses because they have different numbers of neutrons in the nucleus.

Lagomorphs An order of placental mammals that includes rabbits, hares, and pikas.

Larvae Self-sustaining, immature organisms that for a period after hatching are morphologically dissimilar than adults of the same species.

Lepospondyls A group of Late Paleozoic amphibians (Subclass Lepospondyli) with spool-shaped vertebrae.

Limestone A sedimentary rock composed largely of calcium carbonate in the mineral form of calcite.

Lissamphibia The modern amphibians, including frogs, toads, salamanders and newts.

Litopterns South American ungulates that closely paralleled the horses and camels in their evolution.

Lophophore A circular, half-circle, or coiled ridge surrounding the mouth and bearing slender hollow tentacles with ciliated surfaces; as in bryozoans and brachipods.

Lumbar Pertaining to or situated near the loins (L. *lumbus,* loin).

Lycopsids Members of the Lycopsida; leafy plants with simple closely spaced leaves bearing sporangia on their upper surfaces. Represented by ancient scale trees and modern club mosses.

Mantle The zone in the earth's interior below the crust and above the core. In zoology, the fleshy fold that secretes the shell of a mollusc or brachiopod.

Marsupials Mammals of the Marsupialia or Metatheria which have mammary glands but no placenta (or a temporary one), and a pouch or marsupium.

Medusoid Referring to the jellyfish type of coelenterate.

Meiosis The process of cell divsion by which chromosome number is halved during the formation of sperm and egg cells.

Mesenchyme Loosely arranged cells of irregular shape which are relatively undifferentiated and give rise to different tissues.

Mesoblast The layer of cells in the gastrula lying between the outer epiblast and the inner hypoblast.

Metabolism The processes in an organism by which protoplasm is produced, maintained, and destroyed; and by which energy is made available for the functioning of the organism.

Metazoan A multicellular animal.

Miacids An extinct group of terrestrial flesh eaters of the Order Carnivora.

Micron A unit of linear measure; one-thousandth of a millimeter.

Mitosis The usual method of cell division characterized by exact duplication of chromosomes. Mitosis involves the longitudinal splitting of each chromosome and the distribution of one half of each chromosome to each daughter cell.

Mold The imprint of an organism or a part of an organism in the enclosing or surrounding sediment.

Molecule The smallest particle of an element or a chemical compound that retains the characteristics of the original substance.

Monotremes Egg-laying mammals (Monotremata or Prototheria) that have mammae of a different type than those in other mammals and retain several reptilian features.

Mosasaurs Large, marine lizards of the Late Cretaceous.

Mustelids Such members of the Order Carnivora as weasels, otters, skunks, martens, and minks.

Mutation A sudden departure from the parent type, as when an individual differs from its parents in one or more inheritable traits as the result of a change in a gene or chromosome.

Natal Pertaining to or dating from the birth of an individual.

Natural Selection The selection of organisms for superior viability and reproductive success under natural conditions, resulting from the tendency to increase in numbers by reproduction and the limitation of this increase by elements in the environment.

Nautiloids Cephalopods which typically have straight to increasingly coiled external conchs, with subcentral siphuncles, and sutures in straight or gently curved patterns.

Nematocyst A minute stinging capsule with which coelenterates such as jellyfish may inject a paralyzing toxin into their prey or enemies.

Neural Arches Bony or cartilaginous arches formed on the dorsal side of vertebrae above the centrum so as to enclose the spinal cord.

Nothosaurs Early Mesozoic, relatively small sauropterygians. The nothosaurs were less specialized than the plesiosaurs that replaced them during the Jurassic Period.

Notochord A rodlike cord of cartilage cells forming the primary axial structure of the chordate body. In vertebrates, it is present in the embryonic stage and later supplanted by the vertebral column.

Nuclear Fusion That process in which two light nuclei unite to form a heavier one. Considerable energy (thermonuclear power) is evolved in the process.

Nucleic Acids Any of a group of organic acids that control the processes of heredity within cells and the manufacture of protein from the amino acids ingested by the cells as food.

Occipital Condyle The rounded knob or knobs of bone by which the skull is attached to the first vertebra.

Olfactory Pertaining to the sense of smell.

Omnivorous Feeding upon both plants and animals.

Opal A form of amorphous silicon dioxide containing up to 10 percent water. Commonly replaces wood, shell, or bone.

Organelle A specialized structure having a definite function in a cell.

Ossification The formation of bone.

Ostracoderms Extinct, typically armored, jawless "fishes" of the Class Agnatha.

Oviparous Egg laying; producing young which develop from eggs that hatch outside of the body of the mother.

Ovoviviparous Producing young which develop from eggs that hatch within the body of the mother.

Ozone A form of oxygen whose molecules consist of three oxygen atoms each.

Paedomorphosis Transformation whereby the immature or larval form of an organism becomes capable of reproduction.

Palatoquadrate The upper jaw element that extends forward and backward, below and beside the braincase.

Palaeospondylids Small enigmatic placoderms of the genus *Palaeospondylus*, known from numerous specimens found in the Old Red Sandstone of Scotland.

Pangaea Name applied by Alfred Wegner to a vast, hypothetical, Late Paleozoic supercontinent from which all others originated as a result of fragmentation and drifting.

Peccaries Relatively small artiodactyls of the pig family. As compared to pigs, peccaries are smaller, predominantly two-toed, have longer legs relative to their bodies, and have shorter deeper skulls.

Pecoran The advanced ruminants, including deer, giraffes, sheep, antelopes, and cattle.

Pectoral Pertaining to the anterior or upper thoracic region, as in pectoral fins, pectoral girdle, or pectoral limbs.

Pedicle In brachiopods, a fleshy stalk by which the brachiopod may be attached to the seafloor.

Pelagic Of or inhabiting the open water, away from shore. The term usually applies to the open ocean environment.

Pelvic Girdle The group of bones to which the bones of the hindlimb are attached, connecting the hindlimbs to the axial skeleton in the vertebrates.

Pelycosaurs Early mammallike reptiles exemplified by the sail-back forms of the Permian Period.

Period A subdivison of an era in the geologic time scale.

Perissodactyls Those progressive, hoofed mammals which characteristically have an odd number of toes on the hind feet and, usually, the front feet.

Phanerozoic Eon The eon of geologic time characterized by the existence of abundant and diverse life. The Phanerozoic follows the Cryptozoic and is divided into Paleozoic, Mesozoic, and Cenozoic Eras.

Phenotype The expressed traits of an individual as contrasted with its genetic constitution or genotype. The sum total of the realized physiological and structural traits of an individual.

Photosynthesis The biological process by which plants transform carbon dioxide and water into carbohydrates by using solar energy absorbed by chlorophyll.

Phthinosuchians Mammallike reptiles intermediate in morphological development between pelycosaurs and therapsids.

Phytoplankton Marine planktonic unicellular plants such as diatoms, microscopic algae, and dinoflagellates.

Pineal Gland A structure located on the roof of the brain of vertebrate animals; phylogenetically associated with a median eye.

Pinnepeds Marine carnivores such as seals, sea lions, and walruses.

Pistil The seed-bearing organ of a flower. The complete pistil consists of ovary, stigma, and style.

Placenta The organ by means of which the placental mammalian embryo is nourished during development. The placenta is derived in part from the uterine wall and in part from the embryonic membranes.

Placoderms Extinct, primitive jawed fishes of the Paleozoic Era.

Placoid Scales Small, conical or toothlike scales such as those found in sharks, rays, and their relatives.

Plagiosaurs Grotesque Late Triassic amphibians characterized by broad, flat heads and bodies, and small limbs.

Plankton The term which collectively refers to organisms that float or drift at random in a body of water.

Plastron (of turtle) The flat slablike belly structure possessed by turtles, typically consisting of four paired plates with an anterior median element.

Plate Tectonics A concept that the earth's crust and upper mantle can be divided into huge moving plates, with volcanism and mountain building associated with specfic boundary conditions between plates.

Pleuracentrum (of vertebra) Wedge-shaped element posterior to the centrum in the vertebrae of labyrinthodont amphibians.

Pollen The male fertilizing elements (microspores) of flowering plants.

Polymer A compound, usually of high molecular weight, formed by linking of simpler molecules.

Polyp The attached hydralike form of some coelenterates.

Postorbital Bone One of the series of bones that surround the orbital opening, located in the dorso-posterior position.

Premolar Mammalian teeth located posterior to the canines and in front of the molars. Premolars are the "bicuspids" of dentists.

Proboscis In reference to tapirs, an elongated flexible snout; in elephants, a long, prehensile extension of the nose and upper lip; and in acorn worms, the protruding front of the "head."

Procaryotes Cells which lack a true nucleus, having nuclear materials within the cytoplasm.

Prograde Said of a planetary body which is rotating in the same general direction as its direction of revolution.

Prosimii The less advanced primates; lemurs, tarsiers, and tree shrews.

Proteinoids Extra large molecules containing most of the twenty amino acids found in proteins and produced in the laboratory under conditions simulating those to be found on earth.

Proteins	Any of a group of complex organic nitrogenous compounds widely distributed in plants and animals and required for all life processes. Proteins are polymers of amino acids.
Protista	A kingdom of organisms which includes those organisms transitory or intermediate between the Plant Kingdom and Animal Kingdom.
Protorosaurs	Ancestral, terrestrial euryapsid reptiles of the Permian and Triassic.
Pseudopodia	Temporary projections from the membrane of a cell that are involved in both feeding and locomotion.
Psilopsids	Generally leafless, rootless, simply constructed plants with sporangia located on the terminations of the axes.
Pterosaurs	Extinct diapsid reptiles modified for flight.
Pubis	The more anteriorly located of the bones in the lower part of the pelvic girdle.
Pygidium (as in trilobites)	The posterior portion of a trilobite, consisting of two or more coalesced segments.
Quadrate Bone	In amphibans and most reptiles, the quadrate bone forms the articular surface for the lower jaw.
Quartzite	A metamorphic rock formed of intergrown crystals of quartz and typically resulting from the metamorphism of quartz sandstone.
Radius	One of the two long bones of the lower part of the limb. In man, the shorter bone of the forearm on the thumb side.
Respiration	The delivery of oxygen to the protoplasm of cells, either directly by diffusion from a surrounding fluid (as in protozoa and simple metazoa) or indirectly by means of organs of respiration and distributing systems.
Rhachitomes	Labyrinthodont amphibians of the Late Paleozoic, of which *Eryops* is a well-known example.
Rhenanids	"Sharklike" placoderms of which *Gemuendina* is a well-known representative.
Rhynchocephalian	Diapsid reptiles represented by one living species, *Sphenodon punctatus,* which is restricted to a few islands off the coast of New Zealand.
Rudimentary	Incompletely developed or having no function.
Rugose Corals	Exclusively Paleozoic solitary or colonial corals with septa arranged in cycles of four.
Rumen	Large organ in grass-eating mammals where food is stored

and fermented, and from which this food is regurgitated and rechewed before traveling to the stomach.

Ruminant A herbivorous land mammal that chews a cud, as a deer or cow.

Sacral Pertaining to the posterior part of the vertebral column that is attached to the pelvic girdle.

Sarcodinids Unicellular members of the Phylum (or Subphylum) Sarcodina. Sarcodinids are eucaryotes and may be either naked or secrete encasements, as in radiolarians or foraminifers.

Sauropod Large, herbivorous, saurischian "dinosaurs."

Sauropterygians Marine reptiles of the Mesozoic, including the well-known plesiosaurs.

Sciuromorphs The squirrels, chipmunks, and marmots.

Scleractinian Corals The most numerous and varied of the modern anthozoan corals. Scleractinians include the stony hexacorals of modern seas and have septa arranged in cycles of six.

Seafloor Spreading The mechanism by which new oceanic crust is created along lines of crustal divergence (midoceanic ridges) and adjacent tectonic plates are shifted laterally to provide space.

Sedentary Pertaining to organisms that are more or less permanently located or attached to a single location, as an oyster or a coral polyp.

Sedimentary Rocks Rocks formed by the deposition of sediment (sand, clay, lime) and its consolidation by compaction or recrystallization.

Septum A partition between two chambers or cavities.

Series The rocks deposited or emplaced during the span of a geologic epoch.

Sessile Sedentary or attached in one place.

Silicate Any of a large class of minerals containing silicon and oxygen, and constructed from the tetrahedral group $(SiO_4)^{-4}$. (Most of the earth's crust is made up of silicate minerals.)

Siliceous Containing silica, SiO_2.

Sirenians Sea cows, such as manatees and dugongs.

Species The major subdivision of a genus or subgenus; composed of related individuals capable of interbreedng among themselves.

Sphenopsids The plants referred to as articulates or horsetails, with branches and leaves carried in whorls.

Spiracle A small vent or opening. In blastoids, spiracles are openings allowing the escape of water from the water vascular system. In fishes, the spiracle is an evolutionary modification of the first gill slit.

Spores Asexual reproductive bodies of lower organisms, such as bacteria, ferns, and mosses. Spores are capable of developing directly into an adult.

Sporophyte The asexual diploid generation of a plant that produces spores. (It alternates with the sexual, haploid gametophyte generation.)

Spreading Center Linear areas of the earth's crust where tectonic plates are moving apart as new crustal material is added to their trailing borders.

Squamosal Bone One of the temporal bones of the skull, typically lying posterior to the postorbital.

Stamen The pollen-bearing or male organ of a flower.

Strata Layered sedimentary rocks.

Stratigraphy The study of the definition, correlation, and interpretation of stratified rocks.

Stereospondyl Amphibians A group of labyrinthodont amphibians that had become more continuously aquatic than the rhachitomes, and although large in size, had secondarily weaker vertebrae and limbs.

Stromatolites Laminated calcareous (or secondarily siliceous) mound or fingerlike deposits built up as the result of the life processes of such simple organisms as blue-green algae.

Subduction Zone An inclined, planar zone descending away from a trench and defined by frequent earthquake activity. Subduction zones are interpreted as the zones of shear between sinking oceanic plates and overriding plates.

Sulcus Pertaining to the longitudinal depression typically located on the pedicle valve (the valve which bears the pedicle opening) of a brachiopod.

Sutures In cephalopods, the lines formed at intersections of septa and the inner surface of the wall of the conch.

Symmetry The reversed repetition of parts around an axis or on opposite sides of any plane, so that mirrored halves result from a separation along the axis or plane of symmetry.

Syndactylous Digits Kind of digits found in marsupials, in which the second and third digit are enclosed in a common membrane.

System The rocks formed in the course of a geologic period. For example, during the Triassic Period the sediments of the Triassic system were deposited or emplaced.

Tabula Horizontal skeletal partitions found in the thecae of Paleozoic tabulate corals.

Thallophytes In general, plants with thallus bodies; that is, not differentiated into stems, roots, and leaves. Examples of thallophytes are bacteria, blue-green algae, fungi, and lichens.

Thecodonts An order of primarily Triassic diapsid reptiles generally considered to be the ancestral archosaurians.

Therapsids The order of advanced mammallike reptiles.

Theriodonts A suborder of the Order Therapsida which includes advanced, carnivorous, mammallike reptiles.

Thorax Generally, that part of the body between the stomach and the head.

Theropods Carnivorous, saurischian "dinosaurs."

Tibia The inner and usually larger of the two long bones of the hind limb between the knee and ankle.

Tracheophytes Green plants with specialized conducting tissue (wood and phloem); vascular plants.

Trochophore Larva A bilaterally symmetrical, pear-shaped larvae encircled by a band of cilia. The trochophore is the larvae of molluscs and annelids.

Tylopod Camels and llamas.

Ulna One of the two long bones of the forelimb, between the "elbow" and "wrist."

Ungulate General term for hoofed animals.

Uraninite A uranium oxide mineral commonly known as pitchblende.

Vagrant Pertaining to organisms that freely move about.

Vascular Pertaining to the circulation of fluids.

Ventral Pertaining to or situated near the surface away from the back; the lower part of an animal or organ; opposite to dorsal.

Vertebrae The series of short bone or cartilage segments making up the vertebral column in vertebrates.

Vestigial Small or degenerate, but representing a structure that at one time was more fully developed or functional.

Viscera A collective term for the internal organs (especially abdominal organs) of an animal.

Viviparous Giving birth to young that develop from eggs within the body of the mother and are nourished from the bloodstream of the mother.

Warm-Blooded Animals

Animals whose body temperature is nearly constant and usually higher than that of the surrounding environment; including birds and mammals.

Zooecium

The calcareous or chitinous skeleton of an individual bryozoan (zooid).

Index